BAD ADVICE

Paul A. Offit, M.D.

BAD ADVICE

*Or Why Celebrities, Politicians,
and Activists Aren't Your Best Source
of Health Information*

COLUMBIA UNIVERSITY PRESS

NEW YORK

Columbia University Press
Publishers Since 1893
New York   Chichester, West Sussex
cup.columbia.edu
Copyright © 2018 Paul A. Offit

Library of Congress Cataloging-in-Publication Data
Names: Offit, Paul A., author.
Title: Bad advice : or why celebrities, politicians, and activists aren't
    your best source of health information / Paul A. Offit, M.D.
Description: New York : Columbia University Press, [2018] | Includes
    bibliographical references and index.
Identifiers: LCCN 2017056425 (print) | LCCN 2018010525 (ebook) |
ISBN 9780231546935 (electronic) | ISBN 9780231186988 (cloth : alk. paper)
Subjects: LCSH: Communication in public health. | Communication
    in medicine. | Health in mass media.
Classification: LCC RA423.2 (ebook) | LCC RA423.2 .O34 2018 (print) |
    DDC 362.101/4—dc23
LC record available at https://lccn.loc.gov/2017056425

∞

Columbia University Press books are printed on permanent
and durable acid-free paper.
Printed in the United States of America

Cover design: Noah Arlow

To my wife Bonnie, who stayed strong after every lawsuit and death threat, and to our children, Will and Emily, who were mercifully oblivious.

We live in a society exquisitely dependent on science and technology in which hardly anyone knows anything about science and technology.

—Carl Sagan

CONTENTS

PROLOGUE: ON BEING NAÏVE

Question: How can you tell the difference between introverted scientists and extroverted scientists?

Answer: When introverted scientists talk to you, they stare down at their shoes. When extroverted scientists talk to you, they stare down at your shoes.

In 1997, an executive producer at a local Fox station in Philadelphia asked me to appear on her show. It was September, back-to-school month, and the producer wanted to talk about vaccines. At the time, I was an associate professor of pediatrics at the Children's Hospital of Philadelphia and had been studying vaccines for years. I thought it would be fun.

When I arrived at the studio the producer explained that the segment would have more energy if, instead of sitting at the desk with the two news anchors, I sat in the newsroom. She directed me to a high, unstable chair that rocked with the slightest movement, a cameraman just a few feet in front of my face. Because I couldn't see or hear the news anchors, who were in a room behind me, the producer gave me an earpiece, which fit badly.

The segment following mine involved a legal comedy-drama premiering on Fox called *Ally McBeal*, starring Calista Flockhart.

Ally McBeal featured several lawyers who wore miniskirts to work. Finding this fashion statement intriguing, the producer had asked four models, all in their early twenties, to wear progressively shorter skirts. Talking animatedly among themselves, the models stood next to me. People in the newsroom buzzed around, talking, joking, laughing; one was screaming. My chair rocked back and forth.

At this point, I wondered whether it was possible to be any more distracted. Where were the clowns, the dancing bears, the jugglers, the exotic animal handler searching for his escaped scorpion? Then my earpiece fell out. When I put it back in, I realized that one of the news anchors was asking me a question. My segment, apparently, had started. "Dr. Offit, could you tell us what vaccines children get, how many they get, and when they get them?"

The actual answer to that question in 1997 would have been

> Children receive a vaccine to prevent hepatitis B virus at birth and then at one and six months of age; the combination diphtheria–tetanus–pertussis vaccine at two, four, six, and fifteen months and again at four years of age; a vaccine to prevent *Haemophilus influenzae* type b at two, four, six, and twelve months of age; a polio vaccine at two, four, and six months and again at four years of age; the combination measles–mumps–rubella vaccine at twelve months and four years of age, and the chicken pox vaccine at twelve months of age.

Apart from setting the field of health communication back about twenty years, there was no way I was going to remember all of that. The better answer would have been

> Children receive several vaccines in the first few years of life to prevent pneumonia, hepatitis, meningitis, and bloodstream infections, among other diseases. Parents should make sure that their children are up to date on their vaccines so that they don't have to suffer these terrible infections.

I didn't give either of these answers. I didn't give the first one because I couldn't. And I didn't give the second one because I was too inexperienced to realize that you don't have to answer the question exactly as asked. Rather I gave a variant of the first answer during which I got lost in the middle, forgetting which vaccines I had already mentioned and stumbling over exactly when they were given. It was pathetic. In fact, it was so pathetic that even the models stopped talking and stared at me sadly. When it was over, the producer escorted me out of the newsroom, told me how much she had appreciated my coming to the studio, and never asked me to appear on her show again.

Since my Fox interview, I've written books about vaccines, antibiotics, alternative medicine, dietary supplements, megavitamins, faith healing, and scientific discoveries gone awry. In addition, between 1980 and 2006, I was part of a team of scientists at the Children's Hospital of Philadelphia that invented a vaccine. As a consequence, I've had many more opportunities to interact with the media. I've been interviewed on national morning and evening news shows, grilled on comedy shows like *The Colbert Report* and *The Daily Show*, collaborated on scientific documentaries on CNN, *Frontline*, and *Nova*, and appeared before congressional subcommittees. I've learned a lot along the way. Now I make a complete ass of myself much less frequently.

One thing I found that I never would have predicted was that I had inadvertently put myself in the crosshairs of powerful forces intent on defeating science: an unholy alliance working against the health of Americans. By standing up against these groups, which include hostile activists and personal injury lawyers, I've received hundreds of pieces of hate mail, been the target of four death threats, and been threatened with three lawsuits. I've also been physically harassed. It's been an education.

My hope is that by reading this book, people will learn from my journey through the obstacle course of the current culture. Because we learn about our health through the opaque prisms of newspapers, magazines, radio, television, movies, activist groups, industry representatives, celebrities, politicians, and the internet, we often fail to understand where the real risks lie. As a consequence, we don't always make the best decisions for our health and the health of our children. This failure to appreciate how culture shapes knowledge will only cause more needless suffering and death—now and for generations to come.

BAD ADVICE

CHAPTER 1

What Science Is—and What It Isn't

Science is a method of organizing our curiosity.

—Tim Minchin, Australian comedian, actor, and musician

Science delivered us out of the Age of Darkness and into the Age of Enlightenment.

Three hundred years ago, graveyards overflowed with small, white coffins. Children died from smallpox, meningitis, pneumonia, whooping cough, bloodstream infections, scarlet fever, diphtheria, hepatitis, measles, and food poisoning. Of every hundred children born, twenty would be dead before their fifth birthday. Mothers died from tuberculosis and childbed fever. Crop failures led to famines and starvation. Homes were infested with filth and vermin. The average life span was thirty-five years.

Scientific advances have eliminated most of this suffering and death. Vaccines, antibiotics, sanitation programs, pest control, synthetic fertilizers, X-rays, air conditioning, recombinant DNA technology, refrigeration, and pasteurization—to name just a few—have allowed us to live longer, better, healthier lives. During the last hundred years alone, the life span of Americans has increased by thirty years.

Unfortunately, scientific discoveries have a darker side. Physicists have given us atomic bombs, which, in 1945, were dropped on civilian populations in Hiroshima and Nagasaki, killing more than one hundred thousand people. Chemists have given us opioids like heroin and fentanyl, which kill about sixty thousand Americans every year. And, through what are called gain-of-function studies, biologists have invented ways to make deadly viruses and bacteria even more contagious; these new microbes have the capacity to cause plagues more devastating than anyone has ever encountered.

The fear that our reach has exceeded our grasp is reflected in movies like *Frankenstein* (1931), *Colossus: The Forbin Project* (1970), *Westworld* (1973), *Jurassic Park* (1993), *28 Days Later* (2002), *Splice* (2009), and *Ex Machina* (2015), each depicting a world in which science is out of control—something to fear, not embrace. "Our scientific power has outrun our spiritual power," said Martin Luther King Jr. "We have guided missiles and misguided men."

· · · ·

LET'S START BY TAKING A STEP BACK AND ASKING A BASIC QUESTION: What is science?

Stripped to its essence, science is simply a method to understand the natural world—it's an antidote to superstition.

In a sense, everyone is a scientist. For example, if a car doesn't start, a mechanic considers several possibilities: the battery is dead; the starter is defective; the car is out of gas; the fuel system is clogged. Then, the mechanic tests each of these potential problems. This is exactly how scientists think: hierarchically, reasonably, logically. It would be unreasonable to think that the car doesn't start because the gods were angry or because the owner had cheated on his taxes. That's not scientific thinking; that's magical thinking. And it's at the heart of a lot of misconceptions about how the world works and whether one thing causes another.

By far the most important part of the scientific process is reproducibility. If a scientist's hypothesis is right, then other investigators will confirm that it's right. If it's wrong, they won't. The beauty of science is that it's enormously self-correcting, questioning, probing, skeptical, mutable. Nothing is a fact until it's reproduced again and again and again.

It will probably come as a surprise to learn what science isn't—it isn't scientists or scientific textbooks or scientific papers or scientific advisory bodies. As once-cherished hypotheses are disproved, scientists throw away their textbooks without remorse. For some people, this is unnerving. We want certainty, especially when it comes to our health. And that's when we get into trouble—what I'll call the "Bones McCoy seduction."

In the long-running television series *Star Trek*, Dr. Leonard H. "Bones" McCoy was the chief medical officer aboard the USS *Enterprise*. To make a diagnosis, McCoy briefly scanned the patient with a hand-held device called a tricorder. He then carefully examined the read-out. And that was it. If the scanner displayed a particular diagnosis, then that was the diagnosis. No questions. No doubts. This kind of certainty is attractive. And it's why doctors like Mehmet Oz and Deepak Chopra, who often represent themselves as all-knowing gurus, are so seductive. They, too, express only certainty. They, too, know the truth, and their truth is immutable, fixed. Unfortunately, medical science doesn't work that way; we'll know much more in a hundred years than we know now. Nevertheless, when our health is at stake, it's hard to accept that our knowledge is incomplete.

But take heart—scientific truths do emerge. Sometimes they emerge over months, sometimes years, and sometimes decades. But they do eventually emerge. And when these truths emerge, they become immutable. Evolution and climate change, for example, are no longer opinions; they're facts built on a mountain of evidence.

Although we should trust the scientific process, we should be skeptical of scientists. Scientists get it wrong all the time. But if they're wrong, they'll eventually be shown to be wrong. There's no hiding. For example:

• In 1926, Johannes Fibiger, a Danish scientist, won the Nobel Prize for his discovery of a worm he called *Spiroptera carcinoma*, which he believed caused cancer. Fibiger was an instant hero. At last, the cause of cancer had been found. But Fibiger was later proven wrong. Worms don't cause cancer.

• In 1935, Egas Moniz, a Portuguese neurologist, won the Nobel Prize for inventing a surgical cure for anxiety, paranoia, schizophrenia, and bipolar disorder. He called his technique a leucotomy; when it crossed the Atlantic Ocean, it was called a lobotomy. The *New York Times* hailed Moniz as a "brave explorer of the human brain." Over the next three decades, more than forty thousand lobotomies were performed across the globe, twenty thousand in the United States alone. But lobotomies didn't cure anything. Rather, they caused memory loss, seizures, and, occasionally, fatal, uncontrollable bleeding. By the 1970s, lobotomies were relegated to the dusty bin of discarded psychiatric therapies, next to whips, chains, and snake pits.

• In 1957, the American physiologist Ancel Keys published a paper claiming that people who consumed less fat had a lower incidence of heart disease, coining the term "heart-healthy diet." Keys was a well-respected scientist, a best-selling author, and a consultant to the World Health Organization and the United Nations. In 1961, he even appeared on the cover of *Time* magazine. When Ancel Keys gave advice, people listened. Because of Keys, margarine, which contained partially hydrogenated vegetable oils, became the "heart-healthy" alternative to butter, which contained animal fats. Although he didn't realize it at the time, Keys had driven Americans into the waiting arms of trans fats. Four decades later, the Harvard School of Public Health estimated that trans fats were causing about two hundred fifty thousand heart-related deaths every year.

- In 1981, after interviewing more than four hundred people, Brian MacMahon concluded that excess coffee drinking increased the risk of pancreatic cancer. MacMahon, a researcher at the Harvard School of Public Health, published his findings in one of the most prestigious medical journals in the world, the *New England Journal of Medicine*. However, other scientists couldn't find what MacMahon had found. And the notion that coffee enthusiasts risked a universally fatal cancer faded away.

- In 1989, Stanley Pons and Martin Fleischmann, nuclear physicists at the University of Utah, made a startling announcement. They claimed that they had created energy in a test tube by fusing two small nuclei to form a larger one. This was big news. Nuclear fusion occurs every day on the sun, the Earth's greatest source of energy. But Pons and Fleischmann had created the sun's energy at room temperature, providing a clean, inexpensive, and limitless source. They called their discovery "cold fusion." Utah legislators were so proud of this homegrown breakthrough that they allocated $4 million to establish the National Cold Fusion Institute on the University of Utah's campus. When more than seventy other research teams failed to find what Pons and Fleischmann had found, the hope of cold fusion died a quiet death. The building that once housed the National Cold Fusion Institute still stands, a literal monument to irreproducible science.

Some people hear stories like this and say, "See! That's why you can't trust science. Science gets it wrong all the time." But they're confusing science with scientists. While scientists might have certain biases—and doggedly stick to those biases—the scientific process prevails. Lobotomies, cold fusion, cancer-causing worms, and margarines loaded with trans fats didn't stand the test of time. In other words, while it is reasonable to be skeptical of scientists, it is unreasonable to be skeptical of the scientific process.

In each of these stories, the scientific method won out. But until it did, the public was misled and confused. The reason was simple. Although scientists claim, correctly, that it's all about the data, scientific data don't speak for themselves. Someone has to speak for them. Of all the lessons I've had to learn, this one has probably been the hardest.

I'll give you a specific example.

A few years ago, Amy Pisani, the executive director of Every Child By Two, a vaccine advocacy group, asked me to speak to Tom Harkin, the popular Democratic senator from Iowa. Harkin had requested that $2 million of the budget of the Centers for Disease Control and Prevention (CDC) be set aside to determine whether vaccines were causing developmental disabilities. Every Child By Two had been founded by the former first lady Rosalynn Carter and by Betty Bumpers, wife of the longtime senator Dale Bumpers. For the previous thirty years, these women had worked tirelessly to ensure that all children in the United States had access to the vaccines that could save their lives. In the wake of Senator Harkin's request, Amy Pisani, Betty Bumpers, and I traveled to Washington, DC, to try to talk him out of it.

At the time, about two dozen studies had already examined the relationship between vaccines and developmental disabilities, including autism. My role in this meeting was to explain the power of these studies—to reassure Senator Harkin that what he was proposing to study had already been studied. Harkin was pleasant and affable, and he asked thoughtful questions. The only interaction during the meeting that unnerved me occurred as we were leaving.

Amy Pisani and Betty Bumpers left before I did, each shaking Senator Harkin's hand. I, too, shook his hand and thanked him. But he didn't let go. While holding my hand, he said that a group of scientists had visited him the previous week and said exactly the opposite of what I had just said. Who was he supposed to believe? I told him that it didn't matter what they said or what I said. The only thing that mattered was what the data showed. And that he should be reassured that the studies supporting the safety of vaccines were well performed and irrefutable. I offered to send him the studies. But Harkin wasn't convinced. He continued to hold my hand in his firm grip. "Can you tell me with confidence that vaccines aren't causing permanent damage to children's brains?" he asked. "Yes," I said. "I can." He continued to stare at me, sizing me up. Was I someone he could trust?

In the end, Senator Harkin never asked Congress for that $2 million. I was glad that a lot of money wouldn't be spent testing something that had already been tested. But in some ways, I felt like we had probably prevailed for the wrong reasons. For Senator Harkin, it seemed as if the issue wasn't determining the relative quality of scientific studies as much as finding scientists who appeared to be trustworthy. Although scientists say that it's always about the data, the fact remains that most people don't have the background to sort out good studies from bad. So they make decisions based on the believability of whoever is doing the talking. This means that charismatic scientists with poor data may be more convincing than awkward scientists with quality data. Appearances win out.

We are, all of us, at the mercy of fringe scientists with winning personalities.

CHAPTER 2

White Mice and Windowless Rooms

It's just a lot of little guys in tweed suits cutting up frogs on foundation grants.

—WOODY ALLEN AS MILES MONROE IN *SLEEPER* (1973)

In 2008, John Porter, a Washington, DC, lawyer and former Republican member of Congress, stood in front of a group of scientists at a meeting of the American Association for the Advancement of Science (AAAS). Channeling General George S. Patton, Porter issued a challenge. "You can sit on your fingers or you can go outside your comfort zone and get into the game and make a difference for science," he said. "Neither we, nor the AAAS, nor any other group can do it for you. Science needs you. Your country needs you. America needs you . . . fighting for science!" According to Porter, the time had come for scientists to spread out across the country to explain what they were doing and why they were doing it—to make their case to the media and to the people.

Although scientists are probably in the best position to explain science to the public, several factors are working against them—and they're not trivial: specifically, their training, their personalities, and how they and their work are perceived by the public.

. . . .

IN 2003, WHILE STUDYING FOR HIS ROLE AS A GEOPHYSICIST IN THE movie *The Core*, the actor Aaron Eckhart spent time with several real geologists. To his surprise, they didn't seem much different from anyone else. Eckhart noted that scientists were "just as concerned as you or I about everyday things." Nonetheless, the stereotypical images that many people have of scientists—as portrayed on television shows like *The Big Bang Theory* and movies like *Back to the Future*—isn't that far from the truth. Scientists are often shy, quiet, introverted, and thoughtful—far more comfortable working in isolation than carousing in public. You don't see scientists doing stand-up comedy, appearing on reality television shows, or screaming shirtless in subfreezing weather at football games.

The public's perception of scientists is consistent and ingrained.

In 1957, the anthropologist Margaret Mead asked thirty-five thousand American high school students to complete the following sentence: "When I think of a scientist, I think of . . ." They wrote, "The scientist is a man who wears a white coat and works in a laboratory. He is elderly or middle-aged and wears glasses. He may wear a beard. He may be unshaven and unkempt. He may be stooped and tired. He is surrounded by equipment: test tubes, Bunsen burners, flasks and bottles, and weird machines with dials. He spends his days doing experiments. He pours chemicals from one test tube to another. He scans the heavens through a telescope. He peers raptly through a microscope. He experiments with plants and animals, cutting them apart. He injects serum into animals. He writes neatly in black notebooks."

Twenty-five years later, in 1982, an Australian educator named David Chambers asked forty-eight hundred elementary school students to draw a scientist. In each case, the scientist wore a white lab coat, had unkempt, tousled hair, peered out from behind thick, dark-rimmed glasses, and was male.

Why were these children so uniform in their responses? Where do these images come from? In his book *Mad, Bad, and Dangerous? The Scientist and the Cinema*, Christopher Frayling explains their origins.

The wild hair, Frayling argues, comes from the world's most famous scientist: Albert Einstein, the human symbol of genius. Although other iconic scientists like Archimedes, Marie Curie, Charles Darwin, Galileo, Isaac Newton, Louis Pasteur, Linus Pauling, James Watson, or Francis Crick equally could have been revered, Einstein's universal popularity lies in his simple and easy-to-remember formula: $e = mc^2$. Albert Einstein is so famous that his face has become a cultural icon. His wise and sympathetic eyes appear on E.T. in *E.T. the Extraterrestrial* (1982), his forehead on Yoda in *Star Wars* (1977), and his wild hair on Dr. Emmett Brown (played by the actor Christopher Lloyd) in *Back to the Future* (1985).

The white lab coat, writes Frayling, is a "symbol of neutrality, cleanliness, separation from the rest of the world, and standards—usually male—of professionalism." Frayling also proposes a more ominous meaning. In the mid-1960s, the social psychologist Stanley Milgram—attempting to understand the horrors of Nazi Germany—found that people were more likely to submit to authority when the person running the experiment was wearing a white lab coat. (When I am filmed in my laboratory, producers invariably ask me to put on a white lab coat, which I never wear. This, presumably, is to make me look like I actually know what I'm talking about.)

Thick, black glasses are also part of the uniform. Dr. Clayton Forrester (Gene Barry) in the original *War of the Worlds* (1953), the biologist Diane Farrow (Sandra Bullock) in *Love Potion Number 9* (1992), and the MIT graduate David Levinson (Jeff Goldblum) in *Independence Day* (1996) all wear thick, black glasses. Coke-bottle glasses, according to Frayling, "are often an outward and visible sign of the scientist's perceived incompleteness as a human being, a shortsightedness that cuts him or her off from the mainstream."

In short, we perceive scientists as brilliant in the laboratory but unfit to navigate the real world. In *Independence Day*, the wild-eyed, strange-haired Dr. Brakish Okun (Brent Spiner), the director of research in Area 51, meets President Thomas J. Whitmore (Bill Pullman). "Mr. President!" he says. "Wow! This is . . . what a pleasure. . . . As you can imagine, they . . . they don't let us out much." In *I.Q.* (1994), Albert Einstein (Walter Matthau) and two other real-life scientists portrayed by actors pal around in a Three-Stooges–like buddy movie for geniuses. Collectively, they can't drive a car or retrieve a badminton birdie from a tree. "Three of the greatest minds in the twentieth century," notes a friend, "and amongst them they can't change a light bulb." Most pathetic is the pick-up line of the Princeton mathematician John Nash (Russell Crowe) in *A Beautiful Mind* (2001): "I don't exactly know what I'm required to say in order for you to have intercourse with me."

In the worst case, the scientist is seen as someone who creates monsters, either literally, like Frankenstein's monster, or figuratively, like genetically modified organisms (GMOs). "In these images of our popular culture," wrote the historian Theodore Roszak, "resides a legitimate public fear of the scientist's stripped down, depersonalized conception of knowledge—a fear that our scientists, well-intentioned and decent men and women all, will go on being titans who create monsters . . . the child of power without spiritual intelligence."

Because only 0.3 percent of Americans are professional scientists, most people have probably never met one. And so the stereotypes persist. We have no idea who scientists are, what they do, or why they do it. One story, and it is no doubt apocryphal, involves Albert Einstein traveling on a train from New York City to his home in Princeton, New Jersey. Einstein is explaining his theory of relativity to a group of journalists gathered around him. An elderly man sitting across the aisle listens carefully to Einstein's descriptions. When the train reaches

Princeton, the man sidles up to Einstein and says, "So tell me, Mr. Einstein. From this you make a living?"

I'm fairly typical of most scientists. I first started working in a scientific laboratory in 1981, studying a virus called rotavirus: a common cause of fever, vomiting, and diarrhea in young children. At the time, every year in the United States about four million children were infected with the virus, seventy thousand were hospitalized with severe dehydration, and sixty died from the disease. Because rotavirus had only recently been shown to be a cause of human disease, not much was known about how to prevent it. Our laboratory was the first to develop a small-animal model using mice to study this infection. For the next twenty years, every morning I walked into a small, concrete-blocked, windowless room in the animal facility at the Wistar Institute in Philadelphia to inoculate mice and collect their blood, breast milk, and feces. Listening to classical music, I would spend several hours a day, seven days a week, in the "mouse house." As you can imagine, talking to mice every morning alone in a tiny room wasn't the best way to prepare for appearances on *The Colbert Report*.

Indeed, nothing about my job requires me to be good with people. On the contrary, it selects for someone who is perfectly comfortable being apart from people—or at most, working next to them like a child engaged in parallel play. No schmoozing. No backslapping. No gathering around the coffee machine to tell interesting stories from the night before. The opposite of a "people person."

In fact, scientists are so reticent to appear in public that they are often appalled when other scientists do it. They feel that these "celebrity" scientists, by pandering to the media, are selling out; that, by simplifying their work for the public, they're lessening its importance. Perhaps no two people have been punished more for their frequent media appearances than Carl Sagan, whose award-winning 1980 television series, *Cosmos*, sparked an interest in astronomy among thousands of young people, and Jonas Salk, the inventor of the first polio vaccine and one of the first scientists to appear on television. Members of the National Academy of Sciences—one of the most prestigious scientific organizations in the world—refused admission to both Sagan and Salk because of their celebrity. Surely, no *real* scientists would prostitute themselves by doing what they had done.

Another force working against scientists is their training. Early on, scientists learn that the scientific method doesn't allow for absolute certainty. When scientists formulate a hypothesis, it's always framed in the negative; this is known as the null hypothesis. When communicating science to the public, the null hypothesis can be a problem.

I'll give you an example. Suppose you want to know whether the measles–mumps–rubella (MMR) vaccine causes autism. The null hypothesis would be "the MMR vaccine does not cause autism." Studies designed to answer this question can result in two possible outcomes. Findings can reject the null hypothesis, meaning that autism following the MMR vaccine occurs at a level *greater* than would be expected by chance alone. Or, findings cannot reject the null hypothesis, meaning that autism following the MMR vaccine occurs at a level *expected* by chance alone. The temptation in the first case would be to say that the MMR vaccine causes autism and in the second that it doesn't. But scientists can't make either of those statements. They can only say that one thing is associated with another *at a certain level of statistical probability.*

Also, scientists can never *accept* the null hypothesis; said another way, *they can never prove never.* Brian Strom, formerly the head of the Center for Clinical Epidemiology and Biostatistics at the University of Pennsylvania, used to call it "the P word." He wouldn't let his trainees say *prove* because epidemiological studies don't prove anything. When trying to reassure people that a particular health scare is ill founded, the scientific method can handcuff scientists.

Here are some practical examples of Brian Strom's "P-word" problem. When I was a little boy, I watched the television show

Adventures of Superman, starring George Reeves. One thing that any child watching that show knew to be true was that Superman could fly. When you're five years old, television does not lie. I believed that if I walked into my backyard, tied a towel around my neck (to simulate Superman's cape), and jumped from a chair, I could fly. After several attempts (spoiler alert), I found that I couldn't. But this didn't prove that I couldn't fly. I could have tried a million times, and that still wouldn't have proved that I couldn't fly. It would only have made it all the more statistically unlikely. You can't prove that weapons of mass destruction weren't hidden somewhere in Iraq; you can only say that they weren't anywhere that you looked. You can't prove that I've never been to Juneau, Alaska (even though I've never been to Juneau, Alaska); you can only show a series of pictures of buildings in Juneau with me not standing next to them. Scientists know that you can never prove never. The point being that, unlike mathematical theorems, when it comes to studies designed to determine whether one thing causes another, there are no formal proofs—only statistical associations of various strengths.

One example of how the scientific method can enslave scientists occurred in front of the House of Representatives Committee on Government Reform. On April 6, 2000, a Republican member of Congress from Indiana, Dan Burton, certain that the MMR vaccine had caused his grandson's autism, held a hearing to air his ill-founded belief. At the time, one study had already shown that children who had received the MMR vaccine had the same risk of autism as those who hadn't received it. (Since that hearing, sixteen additional studies have found the same thing.) The scientists who testified at the hearing, however, knew that no scientific study could ever *prove* that the MMR vaccine does not cause autism. They knew they could never say, "The MMR vaccine doesn't cause autism." So they didn't. Rather, they said things like, "All the evidence to date doesn't support the hypothesis that the MMR vaccine causes autism." To Dan Burton, this sounded like a door was being left open—like the scientists were waffling or worse, covering something up. "You put out a report to the people of this country that [the MMR vaccine] doesn't cause autism, and then you've got an out in the back of the thing," he screamed. "You can't tell me under oath that there is no causal link, because you just don't know, do you?"

Another force working against scientists is the difficulty of reducing complex scientific issues into simple sound bites. A scientist's instinct is to fully explain an issue—including the nuances and complicated parts—so that ambiguity can be reduced or eliminated. Trying to condense a difficult concept into a sentence or two not only feels intellectually dishonest, it *is* intellectually dishonest.

The "sound-bite" problem is impossible to avoid. For example, in the late 1990s parents became concerned that a mercury-containing preservative in vaccines called thimerosal might cause autism. At the time, thimerosal was present in several vaccines given to infants and young children. The preservative was used in multi-dose vials (which typically contain ten doses) to prevent contamination with bacteria or fungi that might have been inadvertently injected into the vial while removing the first few doses. (Before preservatives were added to vaccines, these bacteria and fungi occasionally caused severe or fatal infections.) Exercising caution, the Public Health Service urged vaccine makers to take thimerosal out of vaccines and switch to single-dose vials, which would eliminate the need for a preservative. This would make vaccines more expensive—about 60 percent of the cost of a vaccine is its packaging—but better safe than sorry.

This wasn't a trivial issue. The firestorm created by the Public Health Service's directive to remove thimerosal from vaccines drew international media attention and gave birth to at least three anti-vaccine groups: Generation Rescue, which believed that children with autism could be cured by removing mercury from their bodies; SafeMinds, which believed that the symptoms of autism were identical to those of mercury poisoning;

and Moms Against Mercury, which advocated for mercury-free vaccines. Mercury is never going to sound good. There will never be an advocacy group called the National Association for the Appreciation of Heavy Metals standing up in defense of mercury. Frankly, any parent could reasonably conclude that if large quantities of mercury can damage the brain (which they can), then even small amounts like those once contained in several vaccines should be avoided.

The media had questions. Why was mercury being injected into babies? And why had the federal government allowed it to happen? Unfortunately, few scientists stepped up to provide answers. And it's easy to understand why. Trying to explain briefly that thimerosal in vaccines had never caused a problem—and why it never would have caused a problem—was nearly impossible. To fully explain this issue, a scientist would have had to have made the following points: (1) Thimerosal, the preservative in vaccines, is ethylmercury; (2) environmental mercury, which is the kind of mercury that can be harmful, is methylmercury; (3) ethylmercury and methylmercury are different (at this point, you've already lost the audience; prefixes like *ethyl* and *methyl* are meaningless; it sounds like you're talking about gasoline or alcohol, which also seem like things that shouldn't be injected into babies); (4) ethylmercury (thimerosal) is eliminated from the body ten times faster than methylmercury, which is one of the reasons that thimerosal doesn't cause harm; (5) methylmercury is present in everything made from water on this planet, including infant formula and breast milk; (6) the quantity of methylmercury in infant formula and breast milk is *much greater* than the trace quantities of ethylmercury that were contained in vaccines; (7) because methylmercury is ubiquitous in the environment, all children, including newborn babies, have methylmercury in their bloodstreams; (8) children injected with ethylmercury (thimerosal)-containing vaccines have levels of mercury in their bloodstreams well within those considered to be safe; and (9) seven studies have shown that children who had received thimerosal-containing vaccines were no more likely to develop mercury poisoning, autism, or other developmental problems than children who had received the same vaccines without thimerosal. Try to reduce that into a ten- or fifteen-second sound bite for television.

I struggled mightily with the thimerosal issue when speaking with the media. I would say things like, "Babies are receiving much greater quantities of mercury from breast milk or infant formula than from vaccines." This wasn't particularly reassuring. Now parents were scared of everything they were putting into their babies—trapped in an environmental hell. Or I would say, "Studies have shown that children who received thimerosal-containing vaccines are at no greater risk of autism than children who received the same vaccines without thimerosal." This, also, wasn't particularly reassuring. From the parents' stand point, mercury is bad, so it shouldn't have been in vaccines in the first place. At one point, I had to testify in front of a hearing during which a member of Congress said, "I have zero tolerance for mercury." What I really wanted to say was, "Mercury is part of the Earth's crust. If you have zero tolerance for mercury, you should move to another planet." (If you've ever testified in front of Congress, you would understand why having representatives move to another planet is not the worst idea.)

In the end, thimerosal was removed as a preservative from virtually all vaccines given to young children, and, because single-dose vials replaced multi-dose vials, the cost of vaccines increased, with no benefit. Advocates perceived vaccines as being safer even though they weren't. They were just more expensive. Scientists who stood up for the science on this issue sounded like they didn't care about children—like they were perfectly willing to stand back and watch babies get injected with a heavy metal. In truth, scientists who stood up for the science of thimerosal *were* standing up for children. Because vaccines were now more expensive, they became less available, putting children at needless risk, especially in the developing world.

Another factor working against scientists is their commitment to precise language; as a consequence, they're intolerant of even the slightest inaccuracies in how science is portrayed to the public. This, in combination with the inevitably inverse relationship between the popularity of a scientific movie or television show and its accuracy, can make for problems. The most dramatic example of this issue is the astrophysicist Neil deGrasse Tyson's reaction to the movie *Titanic*.

Released in 1997, *Titanic*, starring Kate Winslet and Leonardo DiCaprio, was one of the most popular movies ever made, seen by an estimated four hundred million people. And its animated depiction of how and why the RMS *Titanic* sank was perfect. Although people didn't watch the movie because they wanted to learn about the physics of the RMS *Titanic*'s sinking, they learned anyway. (Come for the love story. Stay for the science.) Neil deGrasse Tyson, however, was appalled at the inaccuracy of one scene, so he sent a message to James Cameron, the director. "Neil deGrasse Tyson sent me quite a snarky email saying that, at that time of the year, in that position in the Atlantic in 1912, when Rose is lying on the piece of driftwood and staring up at the stars, that is not the star field she would have seen," said Cameron. To his credit, Cameron revised the scene for the 3D version of the film. One can only imagine that of the hundreds of millions of people who saw *Titanic*, Neil deGrasse Tyson was the only one who watched that scene and thought, "That's not what the sky in the North Atlantic would have looked like at that time of the year in 1912! What the hell was James Cameron thinking?"

My favorite movie mistake is in *The Wizard of Oz* (1939). After the Wizard gives him a diploma, the scarecrow recites the Pythagorean theorem: "The sum of the square roots of any two sides of an isosceles triangle is equal to the square of the remaining side. Oh joy, rapture, I've got a brain!" An isosceles triangle has two equal sides. The square roots of one of those equal sides plus the third side cannot possibly equal the remaining one. He should have said, "The square root of the *longest side* of a *right* triangle is equal to the sum of the squares of the other two sides." (My children argue that it's not fair to complain about the scientific accuracy of a movie that includes a scene in which an entire house taken up by a tornado in Kansas kills a wicked witch in Munchkinland.)

Given all of these issues, scientists might appear to be the last group able to effectively communicate science to the public. They do, however, have one thing going for them. And it's probably going to surprise you. Scientists are good storytellers. They have to be. Federal funds to support science are limited, and it's hard to get scientific papers published in well-respected journals. To survive, scientists must compel people listening to their talks or reading their articles or reviewing their grant proposals that what they are doing is important. Otherwise, they won't be able to do it much longer.

I'll give you a personal example. Many years ago, I published a paper in the *Journal of Virology* titled "Molecular Basis of Rotavirus Virulence: Role of Gene Segment 4." Admittedly, this title won't grab people in the same way that, say, "Disco Bloodbath" would; nonetheless, there was a story to tell. In the first few paragraphs of this and my other scientific papers, the reader would learn that rotaviruses killed about two thousand children *a day* in the world; that when children died, they died from shock caused by severe dehydration; that the World Health Organization, in an attempt to treat a disease that killed more infants and young children than any other single infection, started a program to provide developing-world countries with oral rehydration fluids; that the program was entirely unsuccessful because it's hard to get children to hold down fluids when they're vomiting; that, because of this, the best way to prevent these deaths was with a vaccine; and that our team at the Children's Hospital of Philadelphia had taken a first step in understanding at least one part of this virus that was making children sick. This was a good story, a sort of quest. It wasn't a Homeric quest—no Minotaurs, Trojan Horses, Sirens, or Golden Fleeces. But a quest, nonetheless.

Here's another personal example of how scientists are compelled by story. In March 2016, I was asked to speak at the

Graduate School of Public Health at the University of Pittsburgh. For me, this was a homecoming, as I had done my pediatric training at the Children's Hospital of Pittsburgh in the late
1970s. When I was an intern, I saw a little girl die from a rotavirus infection. The mother, from rural Western Pennsylvania,
was fiercely devoted to her daughter. When the little girl developed fever and began vomiting, she called her doctor, who told
her to give the child small sips of water containing salt and
sugar. Despite her attempts, the child couldn't hold anything
down; the vomiting was just too severe. Later that night, less
than twelve hours after her illness had begun, the child was list-
less. So the mother brought her little girl, only nine months old,
into the emergency department of our hospital. The minute the
mother walked through the door, a nurse whisked the child to
one of the treatment rooms so that we could insert a catheter
into a vein in her arm to give her the fluids she so desperately
needed. Unfortunately, the child was so dehydrated that we
couldn't find a vein. So we called a surgeon to come to the emergency room to thread a catheter into a vein in her neck. While
waiting, we placed a large-bore hollow needle (typically used for
removing bone marrow) into the middle of a bone just below her
knee. We hoped to get enough fluid into her bone, which would
then enter her bloodstream, to prevent the impending shock. But
it was too late. Before the surgeon had arrived, the child's heart
had stopped beating. We tried to resuscitate her, but couldn't.

The next moment was worse than anything you can imagine.
Now, we had to walk out of the treatment room, into the waiting room, and tell a young mother that her nine-month-old
daughter, who was perfectly healthy two days ago, was dead.
That's the story I told when I began my talk in Pittsburgh about
our work on a rotavirus vaccine. And it's the story that was
always in the back of my mind during the twenty-six years that
we developed that vaccine. We are, all of us, compelled by story.

Indeed, scientists have always been good storytellers.

For example, Galileo was the first person to observe the
phases of Venus, the moons around Jupiter, and hundreds of

previously unseen stars in the Milky Way. On March 13, 1610, he published his observations under the title *The Starry Messenger*. What a great title. He could have chosen something far more scientific sounding, like *Planetary Observations*, but he didn't. *The Starry Messenger* suggested not only that he had been observing the heavenly bodies, but that they had been observing him. Later, Galileo angered the Church by claiming that the Earth revolved around the sun, not the other way around. His observation contradicted the biblical statement that the Earth was "the center of the firmament." Galileo was tried by a Church inquisition, found guilty of heresy, and placed under house arrest. Fortunately, we don't do that anymore. We don't arrest scientists and put them in jail when they say things we don't like. Now we just send them threatening emails.

Historically, scientists have also been good entertainers. Electricity, for example, was popularized at county fairs and expositions; people were amazed when they received small shocks, when their hair stood on end, or when they saw a dead frog's leg twitch. Probably no one was more impressed by the frog-leg-twitching demonstration than Mary Shelley, who, at the age of 18 years, wrote a book about how electricity brought a monster to life: *Frankenstein*. (It's alive!)

What Galileo did for astronomy and Mary Shelley did for electricity, Louis Pasteur did for vaccines. In May 1881, Pasteur separated sheep, goats, and cows into two groups. To one group, he gave two shots of what he hoped was a vaccine to prevent anthrax, a common cause of disease in animals and a bane to farmers. To the other group, he gave nothing. Thirty days after the injections, he infected both groups with live anthrax bacteria. All of the animals in the unvaccinated group died, whereas those in the vaccinated group survived. The public was amazed by the result, especially when two of the unvaccinated animals dropped dead in front of them.

Many scientists, past and present, have been wonderful storytellers. Biochemists like Isaac Asimov, oceanographers like Jacques Cousteau, physicians like Siddhartha Mukherjee and Atul Gawande, evolutionary biologists like Stephen Jay Gould and Richard Dawkins, mechanical engineers like Bill Nye, and theoretical physicists like Albert Einstein and Stephen Hawking have all found a way to make difficult concepts accessible and fun.

Scientists willing to enter the world of science communication should also be heartened by the fact that most people actually trust scientists and value what they do. A recent poll asked participants to rate various fields on how much they contribute to society. Topping the list was the military; 84 percent of those polled felt that members of the military made important contributions. Next, were teachers at 77 percent, scientists at 70 percent, medical doctors at 69 percent, engineers at 64 percent, the clergy at 40 percent, journalists at 38 percent, artists at 31 percent, and lawyers at 23 percent. Trailing the field were business executives at 21 percent.

So, as John Porter has urged, scientists need to get in the game. If they read a story in a magazine or newspaper or watch a television program or hear a radio show in which the science is inaccurate, they should call the writer or producer and educate them about their field. It won't change the story that has already been published or aired, but at least those journalists might think again before writing a similar story. And, best-case scenario, scientists might be called upon for their comments the next time the subject is raised. Also, no venue is too small. Scientists should speak at elementary, junior high, and high schools about their areas of expertise—they should become an army of science advocates out to educate the country. Because science is losing its rightful status as a source of truth, now is the time.

Regarding speaking at elementary schools, however, I offer only one piece of advice: Avoid the following scenario.

In 2007, when my daughter was in the eighth grade, her biology teacher asked me to speak to her class about vaccines. I was thrilled. My daughter was mortified. During the drive to school that morning, she spent the entire time trying to convince me not to tell jokes. "Don't tell jokes, Dad. Kids my age won't think you're funny. They don't get old people's jokes. Trust me on this." My daughter attended an all-girls school in suburban Philadelphia. During the talk, the twenty or so girls in her class seemed to be enjoying it. My daughter, on the other hand, stared grimly forward. Never moved. A look on her face that said, "Don't embarrass me in front of my friends. I'm not kidding." It was, without question, the most harrowing talk I've ever given. And I'm including talks in front of congressional subcommittees, talks that have been protested by people marching outside, and talks in front of live national television audiences. The worst-case scenarios in those situations is that people will scream at me, send me hate mail, lobby my hospital's CEO to fire me, or try to get my medical license revoked. But embarrass your daughter in front of her eighth-grade friends and you're a dead man. A walking dead man. (Just for the record, I haven't always been scared of my daughter. It's only been since she learned to speak in complete sentences.)

CHAPTER 3

An Alibi for Ignorance

All I know is what I read in the papers;
that's my alibi for ignorance.

—WILL ROGERS

My attraction to science first came from the weekly television program *Watch Mr. Wizard*, starring Don Herbert. The show ran from 1951 until 1965.

Watch Mr. Wizard was filmed in what appeared to be Don Herbert's garage. Every week one of the neighborhood kids would stop by to learn about science. Herbert didn't wear a lab coat—he worked in shirtsleeves—and his "laboratory equipment" consisted of items commonly found around the house. Using eggs, bottles, straws, and spoons, Mr. Wizard taught us a wealth of scientific concepts. He cut up paper plates to make optical illusions, blew smoke through straws to demonstrate air currents, and used a flashing light to simulate radiation. One of my favorite moments was when he created a vacuum to suck an egg into a bottle; it was like a magic trick.

Mr. Wizard was a phenomenon, launching Mr. Wizard Science Clubs, which more than a hundred thousand children in the United States, Canada, and Mexico joined. I bought the Mr. Wizard science kit, read all of the Mr. Wizard books, and did many of Mr. Wizard's experiments at home. Mr. Wizard made science practical and fun. Anybody could do it.

When Don Herbert died in 2007, one of the first people to applaud his life and work was Bill Nye: "If any of you [have been] happy to learn a few things about science watching *Bill Nye the Science Guy*, keep in mind, it all started with Don Herbert."

Mr. Wizard's show aired at a time when people were more receptive to science and scientific expertise. Today, the audience is not only more skeptical (which is reasonable), but also more cynical (which isn't). In addition, much larger cultural, political, psychological, and religious forces are now at work—as you'll see in the following stories.

We Don't Know Where the Real Risks Lie

My Mr. Wizard moment came when I was asked to speak to a group of reporters and lobbyists in Washington, DC, about the vitamin and supplement industry. For the first and only time in my life, I used Mr. Wizard–like props.

Before the talk, I spent a small fortune buying various dietary supplements at a GNC store. I bought chondroitin sulfate and glucosamine to "promote joint health," concentrated garlic to "lower bad cholesterol," saw palmetto to "shrink prostates," and vitamin E to "support heart health." I wanted the audience to understand that these products weren't what they were claimed to be. Holding up the bottle of chondroitin sulfate and glucosamine, I explained how an excellent study published in the *New England Journal of Medicine* found that this supplement was no better than a sugar pill in treating painful arthritis. I explained how concentrated garlic, which was supposed to lower low-density lipoprotein (bad) cholesterol, had also been studied and found not to work. I explained that although saw palmetto contained a drug that had an anti-testosterone effect—which could help shrink prostates—the amount was far too little to make a difference, as had been shown in two high-quality studies.

I finished by holding up a bottle labeled "Natural E 1000," which contained 3,333 percent of the recommended daily allowance of vitamin E and was supposed to lessen the risk of heart disease, but didn't. Worse, several studies had shown that megadoses of vitamin E actually increased the risk of prostate cancer. I explained that almonds were a rich source of vitamin E. Holding up a large bag of almonds, I said that you would have to eat one thousand almonds to match what was in one gel cap of vitamin E and that we weren't meant to eat one thousand almonds at one time. I argued that swallowing a gel cap of vitamin E was an unnatural thing to do.

I also talked about the safety of these all-too-easy-to-purchase dietary supplements. In 2004, the Food and Drug Administration (FDA) recalled supplements containing the weight-loss product Ephedra because it had caused hundreds of cases of psychosis, hallucinations, paranoia, depression, irregular heartbeat, stroke, and heart attack, as well as at least fifteen deaths.

In July 2013, the FDA recalled vitamin B and C preparations made by the ironically named manufacturer Purity First when its products were found to contain methasterone: an anabolic steroid. The vitamins, which had been made in the same vat that had previously been used to make the steroid, had caused masculinizing symptoms in more than two dozen women in the Northeastern United States.

In October 2013, the FDA recalled a weight-loss product called OxyElite Pro after it had caused more than fifty cases of acute hepatitis and liver failure. One person died, and three required liver transplants.

In December 2013, seven children less than four years of age were admitted to the hospital with severe vitamin D intoxication after being given a fish oil supplement containing four thousand times the quantity listed on the product's label.

In February 2014, the FDA recalled L-citrulline, an amino acid used to treat children with certain genetic and metabolic disorders. The product actually contained a different amino acid.

In September 2014, an herbal cough-and-cold remedy called Bo Ying was found to contain dangerously high levels of lead. The manufacturer had recommended Bo Ying for young children.

In October 2014, an eight-day-old infant born prematurely died in Connecticut's Yale New Haven Hospital after being given a probiotic called ABC Dophilus Powder that had inadvertently contained a mold called *Rhizopus oryzae*. This mold, which is typically found in decaying vegetable matter, was detected in the child's bloodstream as well as in several unopened bottles of the powder at the time of death.

Further, during the past few years, weight-loss products containing sibutramine and potency products containing sildenafil have been pulled from the shelves on an almost weekly basis. Neither of these ingredients was listed on the product label. Sibutramine, which is banned in the United States, has been associated with heart attack and stroke. Sildenafil is Viagra.

I wanted people in the room to understand that the FDA doesn't regulate the dietary supplement industry. When consumers use these supplements, they are doing so at their own risk. One supplement industry lobbyist in the audience pointed out that pharmaceutical products also have severe and occasionally fatal side effects. True enough. But these side effects are invariably listed on the package insert; dietary supplements, on the other hand, because they aren't regulated, don't have package inserts. I argued that, although pharmaceutical products could be harmful, at least parents treating their children's ear infections with 100 milligrams of amoxicillin could be assured that it actually *was* amoxicillin in the package and not Viagra; that it contained 100 milligrams of amoxicillin and not 4,000 or 0 milligrams; that it wasn't contaminated with a dangerous mold; and that it didn't contain undeclared heavy metals.

Mr. Wizard was a wonderful showman. And his use of props allowed him to explain scientific concepts quickly and well. Although I, too, used props during my talk to reporters and lobbyists in Washington, DC, I think it's fair to say that I'm a lot more depressing than Mr. Wizard.

We're Seduced by Celebrity

In November 2004, a reporter from *USA Today* asked me to review a new television series called *House*, starring Hugh Laurie as Dr. Gregory House. After watching the pilot episode, I told the reporter that any doctor who was as abusive, pedantic, obnoxious, and mean-spirited as Dr. House wouldn't last five minutes in a hospital before being called in front of an ethics committee; that the medicine crossed the line from fiction to science fiction; that each of the doctors on the show appeared to be board certified in internal medicine, general surgery, neurosurgery, and gastroenterology (quite a feat); and that doctors in real hospitals weren't that good-looking; if they were, they would have become actors (as the *House* actors had).

House featured attractive doctors because we like to look at attractive people. For example, Jenny McCarthy, Alicia Silverstone, Charlie Sheen, Cindy Crawford, Chuck Norris, Kristen Cavallari, Justin Timberlake, Kirstie Alley, and Aidan Quinn have all been given an opportunity to express their anti-vaccine beliefs in national forums because of their celebrity, not because they have any particular expertise. In truth, the media has been willing to provide a platform for any celebrity who wants to scare parents about vaccines. Anti-vaccine celebrities can be found among sports stars (Jay Cutler and Doug Flutie), politicians (Dan Burton, Bill Posey, and Dave Weldon), and movie stars (Robert De Niro and Rob Schneider). Sometimes just being from a famous family is enough. Which brings us to Robert F. Kennedy Jr.

On July 18, 2014, Keith Kloor, a well-known journalist and instructor at the Arthur L. Carter Journalism Institute at New York University, wrote an article for the *Washington Post* about RFK Jr.'s quest to educate Congress and the public about the supposed dangers of thimerosal—the mercury-containing preservative in vaccines. When Kloor wrote his article, seven studies had already failed to show any relationship between thimerosal in vaccines and autism or any other developmental disorder.

Further, thimerosal had been removed from nearly all vaccines given to young children for more than ten years. Despite the removal of thimerosal from vaccines, the incidence of autism had continued to rise. As one reporter noted, "When your cause goes away and your reputed effect increases, you really need to review your class notes on what cause and effect means." Still, Kennedy marshaled on.

To his credit, Keith Kloor made it clear that the science wasn't on RFK Jr.'s side: "According to the Centers for Disease Control and Prevention, the American Academy of Pediatrics, and the Institute of Medicine, no evidence supports a link between thimerosal and any brain disorders, including autism," he wrote. Nonetheless, Kloor painted Kennedy as a sympathetic figure—a man who dared speak truth to power.

"I know I'm gonna win this one," said Kennedy. "I have the ability to push this over the finish line. I know I do. The truth will prevail."

On March 31, 2016, two years after Keith Kloor wrote his article for the *Washington Post*, I got a chance to meet him at a conference titled "Lost in Translation: Is Science Explained Fairly in the Media?" The meeting, which provided a chance for journalists and scientists to interact, was held in the First Amendment Lounge at the National Press Club in Washington, DC. At times the exchanges were heated (which is why it's good that the Club doesn't have a Second Amendment Room). I asked Kloor why he had given RFK Jr. yet another venue to express a belief that had not only been disproved but could only scare parents away from vaccines.

"As a journalist," he said, "you should talk to everybody. I thought it was a legitimate story. [RFK Jr.] had already talked with a leading candidate from the Democratic party who was running for president, which afforded credence to his claims. So I decided to cover the story."

Keith Kloor is an excellent, thoughtful, compelling journalist. Which is why it really upset me when he said that by talking to a potential presidential candidate, RFK Jr. had found a way to make his claims more valid. How did talking to a presidential candidate change the science? "You're depressing me," I said. "Let's face it. Kennedy has access to the media because he's a son of Camelot. A Kennedy."

In fairness, other science reporters have handled the RFK Jr. story far less sympathetically. Laura Helmuth, writing for *Slate*, wrote, "Kennedy said that he is 'very much pro-vaccine,' and that 'vaccines have saved millions of lives.' They will save even more lives if he and his colleagues stop spreading fear and misinformation about them." Similarly, Jeffrey Kluger, a science writer for *Time* magazine, also refused to fall victim to Kennedy's celebrity. Under the title "RFK Jr. Joins the Anti-vaccine Fringe," Kluger wrote, "Kennedy may deeply believe the rubbish he's peddling—but science doesn't care about your sincerity; it cares about the facts. That doesn't mean he's not in a position to do real harm. Like [Jenny] McCarthy, he has a big soapbox and a loud bullhorn, and every parent he frightens into skipping vaccinations means one more child who is in danger."

We Can't Distinguish Real Experts from Fake Experts

In March 2014, I was asked to speak at the annual meeting of the Association of Health Care Journalists in Denver, Colorado. I talked about a recent outbreak of meningitis at Princeton University. Although several different types of bacteria can cause meningitis, the Princeton outbreak was caused by meningococcus, by far the most frightening.

Meningococcus causes both meningitis, an inflammation of the lining of the brain and spinal cord, and sepsis, an overwhelming bloodstream infection. Although other bacteria can also cause sepsis, meningococcus is unique in the rapidity of infection; a child can be fine one minute and dead a few hours later. Indeed, meningococcus causes such panic in the community that the yearly incidence of the disease could probably be determined simply by reading local newspapers.

In 2013, nine Princeton University students were infected with the same strain of meningococcus. Princeton wasn't alone among college campuses suffering this infection. In 2009, the University of Pennsylvania suffered three cases; in 2011, Lehigh University suffered two cases; in 2013, the University of California, Santa Barbara, suffered five cases; and between 2008 and 2010, Ohio University suffered thirteen cases.

During the Princeton outbreak, the CDC decided to do something that had never been done before—offer an unlicensed medical product to otherwise healthy people. Under compassionate-use protocols, unlicensed medicines are often given to people who are sick. But they had never before been given to people who were healthy. Nonetheless, the CDC decided to offer an unlicensed meningococcal vaccine specific for the strain of bacteria circulating on Princeton's campus to any student who wanted it. This same strain (meningococcus type B) had caused the outbreaks at all of the other colleges.

Many news outlets covered the Princeton story. One was NBC 10 in Philadelphia, which titled its segment "Student Guinea Pigs?" One person interviewed for the story was Dr. Sherri Tenpenny. "I think there's an awful lot that we don't know," she said. "I think that caution is the prudent thing to do here." She urged Princeton students to "do their homework." Otherwise, they might fall victim to a dangerous experiment: "There have been no long-term studies with the meningitis B vaccine to know how long it's going to last, whether or not it's going to work, or what are the long-term potential side effects."

Despite Tenpenny's claims, at the time of the Princeton outbreak, scientists knew a great deal about the meningococcal B vaccine. The vaccine had already been formally tested in more than eight thousand people and had been found to induce a protective immune response, safely. As a consequence, the meningococcal B vaccine had been licensed in Europe, Australia, and Canada and was soon to be licensed in the United States. Nonetheless, by interviewing Sherri Tenpenny, NBC 10 had chosen to scare students about the vaccine, labeling them "student guinea pigs."

Why did NBC 10 pick Sherri Tenpenny to educate its viewers? Tenpenny had never published a single paper about vaccines in a respected medical or scientific journal, never served on a data safety monitoring board, and never been part of a biologics license application. In short, she had no specific expertise on the subject. Furthermore, if the producers at NBC 10 had looked a little closer, they would have realized that Sherri Tenpenny had previously espoused some pretty wacky ideas. In response to a *Huffington Post* article about Adam Lanza—the young man who, after killing his mother, killed twenty children and six adults in an elementary school in Newtown, Connecticut—Tenpenny wrote, "This is a powerful article. . . . I'd like to take one step backward. How much 'mental illness' starts from the sixty-six vaccine antigens and doses of chemicals injected into the developing brain by six months of age?" The implication was that, to Sherri Tenpenny, the killings at Sandy Hook Elementary School weren't caused by Adam Lanza's mental illness or his ready access to guns; they were caused by the vaccines that Lanza had received as an infant. This was the "expert" that NBC 10 had chosen to educate its viewers.

As it turns out, Princeton students *did* do their homework; almost all of the ten thousand students on campus chose to get vaccinated. During this outbreak, I was a member of the meningococcal vaccine working group at the CDC, so I was fortunate to hear periodic updates about the program. The spirit on campus was amazing. Students created and distributed educational materials. They pitched in at vaccination clinics. They published informative articles in the student-run newspaper. And they created beverage holders warning fellow students not to share drinks to avoid spreading meningococcus B bacteria. There was even one moment of comic relief. Regarding the issue of grade inflation at other universities, one Princeton student wore a T-shirt that read, "Would have been meningitis A at Harvard."

At the annual meeting of health care journalists in Denver, I showed the NBC 10 clip featuring Dr. Tenpenny. Then I described some of the journalistic sins that I had witnessed over the previous fifteen years: false balance, little interest in vetting experts, lack of understanding of the scientific method, and the lure of controversy when no scientific controversy existed. If you want to win over an audience, openly criticizing them probably isn't the best place to start. But I was angry that NBC 10 had chosen to scare its viewers, and angrier still at a culture that had encouraged such fear-mongering. (A few months after my talk, several Princeton students visited Drexel University in Philadelphia to attend a party. A nineteen-year-old sophomore named Stephanie Ross, who had been at the party, subsequently died of meningococcal disease caused by the same strain that had been circulating at Princeton. During the weeks that followed, many parents of Drexel students drove up to Canada, purchased the meningococcal B vaccine, stored it on ice, and brought it back to Philadelphia.)

In the question-and-answer session that followed my talk, Dan Diamond, the managing editor of the *Advisory Board Daily Briefing* and a contributor to *Forbes*, asked me whether journalists who wrote misleading stories that caused children to die should be held accountable. Should these journalists be considered party to murder? I proposed journalism jail.

Here's how journalism jail could work. In the NBC 10 example, the producers would have had to appear before a panel of other journalists. If found guilty of producing a show containing inaccurate information that could cause harm, they would be fined $100. And it wouldn't be hard to find a panel of judges. For example, Brian Deer, an investigative reporter for the *Sunday Times of London*; Ron Lin of the *Los Angeles Times*; Anita Manning and Liz Szabo, formerly of *USA Today*; Gardiner Harris of the *New York Times*; Mike Stobbe of the *Associated Press*; Seth Mnookin, who writes for the *New York Times* among other publications; Maryn McKenna of the *Atlanta Journal-Constitution*; Trine Tsouderos, formerly of the *Chicago Tribune*; Maggie Fox from *MSNBC*; Nancy Snyderman, formerly of *NBC*; Amy Wallace, who has written for *Wired*; Tara Haelle from *Forbes*; Helen Branswell from *STAT*; Eula Biss, who has written for *The Atlantic*; and Gary Baum from (of all places) the *Hollywood Reporter*, have all been excellent, hard-nosed, thoughtful, and brave reporters of science and medicine. The money generated from journalism jail could then be put into a scholarship fund for aspiring journalists.

Journalists would argue, correctly, that the First Amendment's right to free speech allows them to unintentionally misinform the public about scientific issues. I agree. That's why journalism jail isn't a real jail.

We Struggle to Distinguish Cause from Effect

A few years ago, I was asked to speak to lawyers from the National Vaccine Injury Compensation Program (VICP) in Washington, DC. The VICP was established in 1986 as part of the National Childhood Vaccine Injury Act. At the time, vaccines were under siege.

In the early 1980s, vaccine makers suffered a series of devastating lawsuits based on the false notion that the pertussis (whooping cough) vaccine had caused cases of permanent brain damage. Even though several studies at the time had shown that the pertussis vaccine didn't cause brain damage, vaccine makers fared poorly in civil court. When the cost of the lawsuits exceeded revenues from the vaccine, several vaccine makers decided to stop making them.

The National Childhood Vaccine Injury Act was put in place to stop the bleeding. Funded by a $0.75 federal excise tax on every dose of vaccine, the VICP, or Vaccine Court, is designed to be a simple one-stop shop for parents who believe that vaccines might have harmed their children. Parents don't have to pay for their own lawyers or experts; the program pays for that. While the VICP compensates people for harms that really were caused by vaccines (like polio caused by the oral polio vaccine or an intestinal blockage caused by an early version of the rotavirus vaccine), it also compensates them for harms that aren't caused by vaccines (like brain damage after receiving a measles vaccine or multiple sclerosis after receiving a hepatitis B vaccine). Still, one could reasonably argue that if the United States is going to mandate vaccines, it's better to overcompensate than to undercompensate for harms, whether real or imagined.

Vaccine Court has a low standard of proof. All a plaintiff must do is (1) show that a vaccine was given shortly before the harm occurred; and (2) propose a mechanism by which the injury could have happened. Also, experts who testify on behalf of plaintiffs in Vaccine Court don't have to prove that they're experts. Almost anyone can claim to be an expert. Although the government, through the Department of Justice, does get a chance to respond, if you can't win in Vaccine Court, you'll have a hard time winning in civil court.

When I was asked to speak in front of the federally appointed judges who ran the VICP and the personal injury lawyers who represented plaintiffs, I wanted to show how ridiculously easy it was to prove cause and effect under this system. So I used an example from my pediatric residency at the Children's Hospital of Pittsburgh. (By the way, speaking in front of a group of about two hundred personal injury lawyers is my concept of Hell, which is what I said when I began my talk. I told the crowd that I was surprised flames weren't lapping up against the walls. For whatever reason, I was the only person in the room who thought this was funny.)

Here's the example I used. When I was a pediatric intern at the Children's Hospital in Pittsburgh, the oncologists diagnosed a five-year-old boy with leukemia. His mother wanted to know what could have caused it. Although some chemicals (like benzene) had been known to cause leukemia, her son hadn't been exposed to any of them. One day she came to the hospital certain that she had figured it out: peanut butter sandwiches. She was sure of it. Her proof? During the previous month, her son had, for the first time, started to eat peanut butter. Wasn't it at least possible that this could have caused his leukemia?

During my talk, I argued that if the VICP was instead the PBICP (Peanut Butter Injury Compensation Program), this mother could have expected to receive about $900,000, the standard settlement. Here's how. First, the child met the criteria for a temporal association; he had eaten peanut butter sandwiches in the month before he was diagnosed with leukemia. Second, it isn't hard to make up a theory for how this could have happened; anyone can make up a theory about anything. Each of the following statements is true: (1) Peanuts are often contaminated with a common fungus called *Aspergillus flavus*; (2) *Aspergillus flavus* makes a toxin called aflatoxin; (3) commercial jars of peanut butter often contain trace quantities of aflatoxin; and (4) large quantities of aflatoxin can cause liver cancer. Therefore, if large quantities of aflatoxin can cause liver cancer, wouldn't it at least be possible for small quantities to cause other cancers, like leukemia? The defense attorney in this case would argue that children who eat peanut butter sandwiches are at no greater risk of leukemia than children who don't eat them. (To my knowledge, this has never been studied, but I don't think I'm going out on a limb here.) Given previous decisions by the VICP, it is likely that the plaintiff's lawyers would have won the peanut-butter-sandwiches-cause-leukemia claim.

The mother who wanted to believe that peanut butter sandwiches might have caused her son's leukemia is fairly typical, normal even. We are always trying to figure out why something happened. If we know why something happened, then we can control it the next time. Unfortunately, we aren't always right. And sometimes we slip into magical thinking, which is far more prevalent than you might imagine. For example, in 2016, a Chapman University survey found that 47 percent of Americans believed in ghosts, 27 percent that aliens have already visited the Earth, 19 percent that people can move objects with their minds, 14 percent that psychics can see into the future, 14 percent that Bigfoot is real, and 40 percent in the existence of advanced, ancient civilizations like Atlantis. Indeed, when she was a little girl, my daughter, Emily, believed in ghosts. When asked why, she said it was because she had seen them on *Celebrity Ghost Stories*. The show featured celebrities such as Tom Arnold, Justine Bateman, Tempestt Bledsoe, Barry Williams, and Cindy Williams, all of whom told personal stories of visitations. I asked Emily why ghosts only visited celebrities at the end of their careers. Unfortunately, this wasn't an easy time for her. She was still getting over the painful realization that the tooth fairy's handwriting was identical to her mother's.

For Emily, magical thinking didn't end with ghosts and tooth fairies. When she was a high school senior, Emily rowed lightweight crew. Before several races, she would occasionally experience tingling in her hands and feet, most likely owing to hyperventilation from nervousness. Her crewmates recommended an asthma inhaler. My wife, looking for something that wouldn't have side effects, offered Himalayan sea salt, which could be purchased on the Internet and inhaled through a porous container. When the package arrived, Emily read the box. Himalayan sea salt was mined from rock formations that were millions of years old! Her eyes lit up. This was going to be great. When I asked her whether she thought that the sodium chloride in Himalayan sea salt differed from the sodium chloride in the saltshaker on our kitchen table, she looked at me and said, "Would you let me believe in something, damn it!" She was right, actually. I should have left that one alone—she really did row better after using the sea salt inhaler.

For me, the best description of our inability to distinguish cause from effect came from a comedian I saw in San Francisco when I was doing research at Stanford. The comedian was pretending to sell a series of Time–Life books, this one titled *The Occult*. He deepened his voice and said, "A woman in California burns her hand on a stove. Her daughter, living three thousand miles away, feels pain in the same hand at the same time! Coincidence?" He waited a beat. "Yes," he said. "That's what coincidence is!"

We Allow Religious Beliefs to Influence
Decisions About Our Health

Human papillomavirus (HPV) is a sexually transmitted infection that causes head, neck, anal, and genital cancers. Currently, every year in the United States, about fourteen million people catch HPV, twenty-nine thousand develop cancers caused by the virus, and five thousand die from those cancers.

The good news is that a vaccine is available to prevent it. In 2006, the committee that advises the CDC, the Advisory Committee on Immunization Practices (ACIP), recommended the HPV vaccine for all adolescent girls between eleven and thirteen years of age; in 2011, the Committee recommended that the HPV vaccine also be given to boys at the same age. Unfortunately, in 2017, only about 45 percent of girls and 40 percent of boys had completed the HPV vaccine series.

One group that made its voice heard loud and clear during the initial ACIP hearings on the HPV vaccine was Focus on the Family, an evangelical organization that rose to prominence in the 1980s. Its stated mission is "nurturing and defending the God-ordained institution of the family and promoting biblical truths worldwide." Focus on the Family promotes creationism, school prayer, and traditional gender roles. Relevant to the HPV vaccine, which prevents a disease that is only spread by sexual contact, Focus on the Family opposes premarital sex and supports abstinence-only sex education. The logic is clear. If people don't have sex before they are married and then remain faithful to their spouse, then neither the husband nor the wife will contract a sexually transmitted disease. However, this type of premarital and marital behavior describes a very small percentage of the American public.

Representatives from Focus on the Family who testified at the ACIP hearings feared that the HPV vaccine would cause teenagers to be sexually promiscuous. For a couple of reasons, this didn't make much sense. First, the HPV vaccine doesn't prevent other sexually transmitted diseases, like chlamydia, gonorrhea, herpes, or syphilis. Indeed, the vaccine doesn't even prevent all strains of HPV, just the ones most likely to cause cancer. In other words, the HPV vaccine isn't a great strategy for preventing sexually transmitted infections, and it's not a contraceptive. Frankly, the logic argued by Focus on the Family was akin to believing that after receiving a tetanus vaccine, teenagers would feel free to run across a bed of rusty nails with impunity.

Nonetheless, Focus on the Family's concern about sexual promiscuity was sufficient to encourage researchers to study it. In 2012, Robert Bednarczyk and colleagues at Emory University in Atlanta examined fourteen hundred girls who either had or hadn't received the HPV vaccine. They found that rates of pregnancy testing, contraceptive counseling, and testing for or diagnosing sexually transmitted infections were the same in both groups. The HPV vaccine hadn't freed teenagers to engage in promiscuous sex.

Three years later, the notion that the HPV vaccine could change sexual behavior reached its illogical end in a standoff between Mike Pence, then the governor of Indiana, and Indiana's health commissioner, Jerome Adams, who would go on to become the United States surgeon general.

During the week of September 21, 2015, the Indiana State Department of Health sent a letter to more than three hundred thousand parents reminding them that their children had yet to receive the HPV vaccine. One of the recipients of this letter was Micah Clark, head of a conservative group called the American Family Association of Indiana. Clark, whose fourteen-year-old daughter hadn't received the vaccine, felt that the letter was intrusive. He asked Governor Pence to look into it. Pence agreed, later asking the Indiana State Department of Health to make it clear that the HPV vaccine was "optional." Pence also reprimanded Jerome Adams for speaking about the HPV vaccine at a state health conference. Later, Pence prohibited the release of a document by the Indiana Cancer Consortium that mentioned cervical cancer, the only known cause of which is HPV.

Ironically, Governor Pence was a vigorous supporter of the meningococcal vaccine, which is given to teenagers at the same time as the HPV vaccine. Given Indiana's population, which is about 2 percent of the nation's, one would have predicted that every year, Indiana's residents would have suffered about ten cases of meningococcus and one death from the disease. HPV, on the other hand, would have caused about sixteen hundred cases of cancer a year and one hundred deaths. So Governor Pence's support of the meningococcal vaccine over the HPV vaccine had nothing to do with the relative health impact of these vaccines and everything to do with his religious ideology. Those suffering from meningococcus infection were innocents. Those suffering from HPV infection weren't.

We're Not Rational Beings

In 2005, I appeared on a New Jersey–based cable news program with host Barry Nolan, who started the show with a surprising question: "If I told you that you had a one-in-one-hundred chance of being killed on the way to the station tonight, would you have shown up for the interview?" Nolan was in New Jersey. I was linked in from an affiliated station in a marginal neighborhood in south Philadelphia at night. Frankly, I think I *had* had a one-in-one-hundred chance of being killed on the way to the station. Nolan was trying to get at the issue of risk. Given that all vaccines have side effects, how big of a risk should we be willing to take?

The notion of how we assess risk has evolved over the last few decades. One of the first scientists to weigh in on the issue was Herbert Simon, who believed that decisions about risk were limited by the availability of information. Simon argued that people will make rational decisions as long as they have the time to consider all of the facts. He called his theory "bounded rationality." In the 1970s, the psychologists Daniel Kahneman and Amos Tversky turned Simon's theory on its ear. They argued that people aren't particularly rational; rather, they form simple rules for evaluating a problem, often ignoring facts that clearly contradict their beliefs. They called these simple rules "heuristics," and believed that they are at the heart of why so many people make bad decisions about their health.

When it comes to making decisions about vaccines, people often resort to what I call the "New York Lottery heuristic."

Let's use the yellow fever vaccine as an example. Yellow fever is a common disease in South America and Africa, causing about two hundred thousand infections a year. Twenty percent of people with severe yellow fever will die from the disease—tens of thousands every year. For these reasons, anyone more than nine months of age who is traveling to areas in South America or Africa where yellow fever occurs should receive the yellow fever vaccine. Some people, however, refuse. Here's why.

In July 2001, the yellow fever vaccine was shown to cause a terrible problem. Some people who got the vaccine developed fever, low blood pressure, respiratory failure, and an inflamed liver—the same symptoms as yellow fever. In other words, the yellow fever vaccine could itself cause yellow fever. Most people who suffered this reaction died. This side effect is extraordinarily rare, with an estimated incidence of one case per ten million doses. In other words, people have a 0.00001 percent chance of experiencing this reaction.

Which brings us to the New York Lottery. To entice people to spend money on the chance of experiencing an extremely rare event, lottery promoters use simple phrases like "It could happen to you" and "Hey, you never know." To prove it, they show pictures of people who won. The chance of winning is as remote as one in fourteen million, roughly the same as getting yellow fever from the yellow fever vaccine. But in both cases, it's possible. It could happen to you. Even though the risk of catching and dying from yellow fever is logarithmically greater than dying from the yellow fever vaccine, some people choose not to get the vaccine, even when traveling to a country where yellow fever is a common, often fatal infection. It doesn't matter whether the risk from the vaccine is one in ten, or one in one hundred, or one in ten million—the fact remains, it's possible.

We're Compelled by Fear More Than Reason

The Ebola virus outbreak in West Africa is a recent example of how easy it is to scare people.

In December 2013, an outbreak of Ebola virus began in Guinea and spread to Liberia and Sierra Leone. Before the outbreak was under control, the World Health Organization estimated that the virus had infected twenty-nine thousand people and caused eleven thousand deaths. One imported case in Dallas led to secondary infections in two medical workers. Also, Craig Spencer, a doctor who had been working in West Africa, was diagnosed with Ebola after he returned home to New York City. When Spencer developed a fever, he immediately quarantined himself and was later admitted to Bellevue Hospital. The day Spencer was admitted, I was in New York City and drove past Bellevue. More than a hundred news trucks were parked outside. People in the city were scared to go to any place that Craig Spencer might have been, including a local bowling alley, for fear of catching the disease.

Ebola is spread by contact with contaminated vomit or feces: bodily fluids that contain large quantities of the virus. Early in the infection, however, when vomiting and diarrhea aren't present, the patient isn't contagious. In other words, if you're sitting next to someone in the early stages of Ebola on an airplane, you're not going to catch it. Although tens of thousands of people were infected with Ebola virus in West Africa, only two people living in the United States caught it. And neither died. As one radio pundit noted on the day that Craig Spencer was admitted to Bellevue, "Americans are at greater risk of being married to Kim Kardashian than of catching Ebola."

When we first made our appearance on Earth, the parts of our brains responsible for fear and emotion—the amygdala and hippocampus—were far better developed than our cerebrum (the thinking part), which was late to the game. In other words, we are hardwired to find food and avoid being food to a far greater extent than we are to dispassionately distinguish coincidence from causality. As a consequence, we're afraid of all the wrong things. We're afraid of vaccines, radon, dioxin, gluten, genetically modified organisms (GMOs), bisphenol A (BPA) and chemicals in toothpaste, deodorant, and shampoo, but not of the things that really can hurt us, like megavitamins, dietary supplements, climate change, and vaccine-preventable diseases.

We Fail to Learn from History

One afternoon, when my son, Will, was five years old, I drove him home from kindergarten. It was President's Week. Will said, "Daddy, I know who the president of the United States is." "Who is it, sweetheart?" I asked. "It's Abraham Lincoln," he declared, proudly. I told him that it actually wasn't Lincoln anymore; it was Bill Clinton. "Bill Clinton?" he said, clearly upset by this new world order. For the next five minutes he stared out the window. I assumed that he was on to other things, but he wasn't. "What happened to Lincoln?" he asked. "Lincoln's dead," I said. "Dead?" he said. "How did he die?" "He was shot," I said. "Lincoln's been shot!?" he screamed.

We all learn from history at different rates. For my son, it took about a hundred and fifty years for an event in 1865 to finally hit him. And when it did, it hit him hard. The same could be said of recent measles outbreaks in the United States.

Before a measles vaccine was first available in 1963, every year about three million children would be infected, forty-eight thousand would be hospitalized, and five hundred would die. Measles was a devastating infection. Thanks to widespread vaccination, measles infections were eliminated from the United States by the year 2000. In 2014, 2016, and 2017, however, measles came back when outbreaks in California, Minnesota, and Ohio affected hundreds of people.

The reason for the outbreaks was that parents, failing to learn from history, had chosen not to vaccinate their children. And the media carried the story with the same wide-eyed disbelief that my son had when he learned that Abraham Lincoln had been shot. It shouldn't be surprising that when enough people choose not to immunize their children, highly contagious diseases can come roaring back to life.

We're Unduly Influenced by Social Media and the Internet

The internet is a source of both great and awful information, allowing people to make either excellent or dangerous decisions for their health. Unfortunately, it's hard to separate reliable websites from misleading ones.

Like it or not, making the best decisions based on the best information boils down to a matter of trust. For example, if something goes wrong with my car, I can visit websites to learn about car mechanics. But no matter how much I read, no matter how many questions I ask friends in chat rooms, the truth is that my best bet is to take the car to someone with experience and expertise—a mechanic. And trust them. I can increase my odds of finding someone who is trustworthy by looking at the list of mechanics recommended on the *Car Talk* website. And, by doing some online research about my car problem, I will be able to ask the mechanic I choose much better questions. But it's unlikely that I'm going to figure out the problem and fix it myself. (As the comedian Jackie Mason said when he opened the hood of his car, "It's so busy in there!")

Vaccines are no different. Sometimes people tell me that after doing their research, they've decided not to get the chicken pox (varicella) vaccine. What they mean by *research* is that they've read other people's opinions about the vaccine on the internet. That's not research. If someone really wants to research the varicella vaccine, they should read the seven hundred papers that have been published on the subject. To do this, they would need to have a working knowledge of virology, immunology, microbiology, statistics, epidemiology, pathogenesis, molecular biology, and clinical medicine. Most people don't have this expertise. In fact, most doctors don't have it. Instead, doctors turn to advisory bodies, like those that report to the CDC and the American Academy of Pediatrics. Collectively, these advisory groups do have that expertise. And collectively, they've read those papers. With this knowledge in hand, these advisory groups make recommendations about who should get the vaccine, who shouldn't get the vaccine, how many doses should be given, when those doses should be given, and what side effects to watch out for.

The current culture, however, does not support the notion that we should trust our experts. Today, many people believe they can know just as much as any expert by Googling topics on the internet or visiting chat rooms. The good news is that people can become much better consumers by reading articles from reliable websites. The bad news is that absent training and experience in a particular field, it's hard to be fully informed. It's also painfully easy to be fully misinformed.

We Allow Our Political Leanings to Influence Our Perception of Science

Anti-science activists were emboldened by the election of Donald J. Trump as president of the United States—and for good reason. In 2014, Trump tweeted, "Healthy young child goes to doctor, gets pumped with massive shot of many vaccines, doesn't feel good and changes—AUTISM. Many such cases!" In August 2016, during a fund-raiser in Florida, Trump met for forty-five minutes with a group of donors that included four anti-vaccine activists. One was Andrew Wakefield, the British researcher who was the first to claim that the MMR vaccine caused autism. According to the participants, Trump expressed a keen interest in holding future meetings with anti-vaccine activists. In January 2017, RFK Jr. claimed that Trump was planning to appoint him to head a commission on vaccine safety and scientific integrity.

Ill-founded concerns about vaccine safety weren't the only anti-science issues supported by the Trump administration.

Climate change denialists were legitimized by the appointments of Scott Pruitt as head of the Environmental Protection

Agency, Rick Perry as head of the Department of Energy, and Sam Clovis as head of the Department of Agriculture. All, like Trump, have denied that human activity has affected the environment.

Evolution denialists were heartened by Trump's choice of Mike Pence as vice-president and—perhaps most dispiriting of all—Betsy DeVos as head of the Department of Education.

Science denialism, however, isn't limited to conservatives. Liberals have waged their own war on science, holding the unshakable beliefs that all things natural are good; anything with a chemical name is bad; and everything that profits an industry is really bad (unless that industry makes megavitamins or dietary supplements).

If you really want to watch science denialism at work, just walk into a Whole Foods store—a beacon in a sea of ill-founded beliefs. At Whole Foods you can forgo one of humankind's greatest scientific advances (by going "GMO free") or be one of the many people to shun a component of wheat that harms only a small percentage of the U.S. population (by choosing gluten-free products).

We'll start with GMOs.

A genetically modified organism is defined as any living organism that possesses "a novel combination of genetic material obtained through the use of modern biotechnology." Genetic engineering, however, isn't a modern phenomenon. Using breeding and artificial selection, humans have been selecting for specific genetic traits in plants, horses, dogs, and other animals for thousands of years. From a practical standpoint, a farmer taking advantage of a chance mutation to cultivate a specific crop a thousand years ago is indistinguishable from a scientist taking advantage of modern GMO technology to create that mutation today. It's still the same mutation.

For farmers, GMO technology has reduced chemical pesticide use, increased crop yields, and enhanced profits, especially in the developing world. In addition, GMO crops have longer shelf lives and better nutrient profiles. And, despite the media hype, GMO foods are as safe as non-GMO foods. The American Association for the Advancement of Science, the National Academy of Medicine, and the European Commission have all issued statements supporting the safety of GMOs. Nonetheless, science denialists continue to damn GMO foods, calling them "Frankenfoods." As a consequence, a recent Gallup poll found that 48 percent of Americans believe that GMO foods pose a serious risk to their health.

Genetic modification has also been used to make life-saving medicines like insulin for people with diabetes, clotting proteins for people with hemophilia, and human growth hormone for children with short stature. Previously, these products were obtained, respectively, from pig pancreases, blood donors, and the pituitaries of dead people. Two cancer-preventing vaccines— which prevent hepatitis B and HPV—are also made using recombinant DNA technology.

The ill-founded fear of recombinant DNA technology reached its illogical end in 2015 when a Democratic member of the New York State Assembly named Thomas J. Abinanti introduced Bill 1706, banning genetically modified vaccines. Abinanti's bill eventually died the embarrassing death it deserved.

In addition to "GMO-free" foods, shoppers at Whole Foods can also find a wonderland of products that are gluten free, even though avoiding gluten benefits only about 1 percent of the United States population.

The gluten story begins in the 1940s.

In 1947, Willem-Karel Dicke, a Dutch pediatrician working in the Hague, described a mysterious condition in which children suffered from diarrhea, bloating, anemia, malabsorption, poor appetite, mouth sores, abdominal pain, and growth failure. Although doctors assumed that the disease was caused by specific foods, no one could identify the offenders. Then something odd happened. At the end of World War II, Holland experienced its *hongerwinter* ("winter of starvation"); many foods, including breads, became unavailable. Although most people in Holland were starving, the children with Dicke's unusual malady thrived.

At a meeting of pediatricians in New York City, Dicke announced that this disease—called celiac disease—was caused by products made from wheat.

In 1953, Dicke identified the specific component of wheat that was causing the problem: gluten, which in some people triggers a destructive autoimmune reaction against the small intestine. Unfortunately, it didn't take long for health hucksters to claim that gluten was causing an impressive list of other diseases.

In 2011, William Davis, a cardiologist, published a book titled *Wheat Belly*. Davis argued that wheat was a modern-day "poison," even though farmers have been harvesting wheat for ten thousand years. Two years later, David Perlmutter, a neurologist, published *Grain Brain*. Davis and Perlmutter claimed that gluten might in part be responsible for an increase in attention deficit disorder, hyperactivity, Alzheimer's disease, arthritis, autism, cancer, heart disease, obesity, and schizophrenia. *Wheat Belly* and *Grain Brain* shared several features: Both were enormously popular; both dramatically expanded the market for gluten-free foods; and both contained no clear proof that their contentions were correct (assuming that you don't count testimonials from Jennifer Aniston, Victoria Beckham, Miley Cyrus, Novak Djokovic, Gwyneth Paltrow, and Oprah Winfrey as proof).

Davis and Perlmutter had ignited a firestorm. Many people with diseases not caused by an autoimmune reaction to gluten now believed that gluten was their problem. Some, who had undiagnosed celiac disease, benefited. (About 85 percent of people with celiac disease in the United States don't know they have it.) Others, who were now eating diets richer in fruits and vegetables, also benefited. Most, however, were doing nothing to improve their health. And they were spending a fortune to do it. A typical basket of gluten-free food costs about three times more than other food.

Today, twenty million Americans claim that they are allergic to gluten when only three million actually are. You can buy gluten-free hair products, gluten-free food for your pets, and go on gluten-free vacations. In 2014, the global market for gluten-free products totaled $4 billion a year; by 2019, it will approach $7 billion.

Marc Vetri, a Philadelphia restaurateur, addressed the issue of gluten-free mania in an article titled "I'm Gluten Intolerant . . . Intolerant." After unsuccessfully trying to explain to a "gluten-intolerant" patron that the risotto she had angrily rejected was made from rice and not wheat, he stopped by the table again at the end of the evening, happy to find her enjoying the rest of her meal. "I offered one last time if there was anything I could get for her. She smiled and said, 'We'll just hang out and let the food digest a bit while we finish our beers.'" Vetri took a moment to admire the irony of the situation. "Sometimes," he wrote, "the jokes are just for me." (Beer contains gluten.)

CHAPTER 4

Feeding the Beast

We learn by going where we have to go.

—THEODORE ROETHKE, AMERICAN POET

When John Porter urged scientists at the American Association for the Advancement of Science in Washington, DC, to get in the game, there was less of a game to get into. By 2008, the number of science articles in American newspapers had shrunk dramatically. The *Boston Globe*, for example, which is located in the heart of the biotechnology industry, had eliminated its science section, choosing to cover health and fitness instead. Science on television had also suffered; less than 2 percent of nightly newscasts included any mention of science. Indeed, the same year that John Porter made his speech, CNN dismissed its entire science, space, and technology unit. Now, only one of every three hundred minutes of cable news is devoted to science.

Nonetheless, occasionally scientists are asked to overcome their personal reticence, the verbal handcuffs imposed by the scientific method, and the cultural, political, and psychological forces working against them to talk to the general public.

What follows are some painful, hard-earned, and occasionally humorous lessons gleaned from personal experience.

· · · ·

IN SEPTEMBER 2007, I WAS ASKED TO APPEAR WITH JENNY MCCARTHY on *Oprah*. McCarthy would be talking about a vaccine she was certain had caused her son's autism. At first, it seemed like a good idea. I could educate millions of people about why vaccines didn't cause autism and how, with so many studies now in hand, we knew that they didn't. A chance to calm the waters.

But that's not what this show was about. Oprah was there to tell a story. And her story had three roles: the hero, the victim, and the villain. Jenny was the hero. Her son was the victim. This left only one role for me. I would be the guy telling Jenny that she was wrong and that, by extension, Oprah was wrong to have had her on the show. Jenny knew the cause of her son's autism (vaccines). I didn't. Jenny had cures for autism (megavitamins, hyperbaric oxygen chambers, and mercury-binding medicines). I didn't. Also, not to be completely politically incorrect, but as a male scientist, you can't go on a television show in front of a studio audience consisting entirely of women and tell two women that they're wrong. It just doesn't work.

Science didn't do well on *Oprah* that day. First, McCarthy told her story: "Right before his MMR shot, I said to the doctor, 'I have a very bad feeling about this shot. This is the autism shot, isn't it?' And he said, 'No, that's ridiculous. It is a mother's desperate attempt to blame something,' and he swore at me, and then the nurse gave [Evan] the shot. And I remember going, 'Oh, God, I hope he's right.' And soon thereafter—boom—the soul's gone from his eyes." Fighting back tears, McCarthy explained how her son had been fine one minute and then, because of that vaccine, had been condemned to a lifelong struggle with autism.

Later in the program, Oprah read a prepared statement from the CDC stating that McCarthy's concerns weren't supported by the evidence—a dry, carefully worded missive from a distant, monolithic body. The CDC didn't have a chance. McCarthy's heartfelt confession won the day.

Jenny McCarthy's appearance on *Oprah* launched her career as a powerful force against vaccines. And we have Oprah Winfrey to thank.

Lesson: Don't go on a show where the host isn't on your side.

・・・・

MANY YEARS AFTER THE FOX INTERVIEW DESCRIBED IN THE prologue, my discomfort during another interview reached the level of parody.

In 2013, I went to a restaurant in southern New Jersey with family and friends. I had been scheduled to participate on a Canadian-based radio show about vaccine mandates. Vaccines are mandated in the United States but not in many other countries, including Canada. In such countries, whether parents choose to vaccinate their children seems to depend more on how citizens view doctors and public health officials than on mandates. In Scandinavian countries, for example, immunization rates are higher than in the United States despite the lack of mandates. Not an easy issue. But I was looking forward to the program. Unfortunately, I'd forgotten about it. Thirty minutes before the show started, the producer called to remind me. I tried to find a quiet spot in the restaurant, but couldn't. So I asked the restaurant manager if there was a room that I could use. She directed me to a business office near the front. Perfect. The office had two rooms separated by a large window. I sat in the back room.

One piece of information that might seem irrelevant to this story, but isn't, is that it was Easter Sunday. When we walked into the restaurant, a large rabbit was handing out candy to the children. This would soon become a problem.

My phone connection was excellent. The host seemed well versed on the issue. And the person who was debating the other side—that vaccines should be optional—wasn't going to argue that vaccines did more harm than good or that I was part of a conspiracy to sell them. I settled in for the hour-long program.

Then the rabbit walked in and sat down in the front room, its back to me. The rabbit didn't see me, but I saw it. Well, not it, really, but her. When she took off her rabbit head, I could see that it was a woman about 25 years old. Then she took off her rabbit suit to reveal that she had nothing on underneath except her underwear.

At this point, I had three options. Option #1: Knock on the window to let her know I was there. But if I did this, I was certain that she would have screamed. Option #2: Don't knock on the window; just hope she doesn't see me. If she were to turn around, however, she would likely scream. Option #3: Turn my back to the window. If she were to see me now, I wouldn't be looking at her, so there would be a chance that she wouldn't scream. I chose option #3. During the entire interview, my heart was in my throat, worried that thousands of Canadians were about to hear a bloodcurdling scream. Screams don't work well for radio shows about science. You want your listeners to be compelled by your arguments, not wondering whether you've just murdered someone in south Jersey.

The rabbit never saw me. When the show was over, I walked past her room, thanked her, and explained what I had been doing. She was surprised but gracious. From this experience, I've learned not to do a radio show when someone in a rabbit suit is undressing in front of me. It's distracting. I suspect that for most scientists this tip will never come in handy.

Lesson: Be comfortable.

. . . .

IN 1993, I HAD MY FIRST TELEVISION EXPERIENCE ON A PROGRAM called *Good Day Philadelphia*. Wally Kennedy was the host. The other guests, Bruce and Whitney Williams, were linked in to the show from their home in Chicago. Whitney, who was eleven years old, had been diagnosed with AIDS the year before. Her father, Bruce, believed that she had acquired the disease from a polio vaccine. While this might sound far-fetched, Bruce Williams wasn't the first person to have made this claim.

The year before the interview, an article published in *Rolling Stone* magazine by Tom Curtis proposed the following series of events: (1) Unknown to investigators, the chimp cells used to make the first lots of polio vaccine contained simian immunodeficiency virus (SIV); (2) after people were injected with the vaccine, SIV mutated into human immunodeficiency virus (HIV); and (3) consistent with this theory, HIV first appeared in the Congo in the 1950s—the same year that a polio vaccine had been tested there. Tom Curtis titled his article "The Origin of AIDS: A Startling New Theory Attempts to Answer the Question 'Was It an Act of God or an Act of Man?' " The article later inspired a book by Edward Hooper titled *The River: A Journey to the Source of HIV and AIDS*. The media loved this story. Now AIDS had a bogeyman: Hilary Koprowski, the scientist who had first tested a polio vaccine in the Congo. According to Curtis and Hooper, by unknowingly infecting children in Africa with SIV, Hilary Koprowski was the father of AIDS.

When I appeared on *Good Day Philadelphia*, highly effective anti-HIV drugs hadn't been invented yet. So AIDS was, for all practical purposes, a death sentence. Whitney Williams would likely be dead in a few years. Her father's confusion and anger were understandable. His daughter had never received a blood transfusion, had never gotten a tattoo, had never used intravenous drugs, and had never, as far as anyone knew, had sex. How in the world had she gotten AIDS? I didn't know the answer. All I knew was that she didn't get it from the polio vaccine. Although it was true that HIV had mutated from SIV in Africa, that event didn't occur in the Congo in the 1950s, as Curtis had claimed; it occurred in Cameroon in the 1930s when, presumably, a hunter inadvertently cut himself while killing a chimp. Also, Koprowski's polio vaccine wasn't made using chimp cells; it was made using monkey cells. Chimps aren't monkeys; they're apes. Finally, and most importantly, in response to the *Rolling Stone* article, researchers went back and tested early lots of Hilary Koprowski's polio vaccine to see whether they contained traces of SIV or HIV. They didn't. In summary, every aspect of Tom

Curtis's proposed series of events was wrong. (Which is one reason *Rolling Stone* isn't considered one of the world's great science journals.)

On *Good Day Philadelphia*, I said that the polio vaccine that Whitney had received had been tested for SIV and HIV and that the viruses weren't there. Unfortunately, I didn't know the source of Whitney's HIV. I only knew what wasn't the source. When Whitney's father got angry with me, I became more insistent. Lost in the discussion, I had failed to express enough sympathy for Whitney.

When I got back to my office that afternoon, my brother, who had watched the show, called me. "Well," he said, noting that I had challenged both Whitney's father and the host, "at least you never attacked the girl."

I felt awful. If I had to do this all over again, I would have said, "Honey, I am so sorry that this has happened to you. I only wish we knew why. All I can say is that it wasn't the polio vaccine. At least we know that much. Hang in there. We're all rooting for you." But I didn't. I was angry that the producer had provided a platform for Bruce Williams's unfounded beliefs and angrier still that the program had used a suffering child to do it. And I let that get the better of me. (What was never revealed on the show was that Bruce Williams was in the midst of suing American Cyanamid, the manufacturer of the polio vaccine.)

On July 18, 1993, a few months after the *Good Day Philadelphia* interview, an article published in the *Chicago Tribune* added a surprising turn to Whitney's story. The article was titled "Mystery of AIDS Girl Has a Twist." Apparently, Bruce Williams had placed two ads in a gay magazine called *Nightlines*. In the first ad, Williams appeared to identify himself as a gay man seeking a lesbian to bear a child (although he later claimed the copy had been rewritten to state that he was gay). The ad read, "I am the parent of a girl who got AIDS from an infected vaccine. My daughter's only wish in life is to be a mother. I want to make her dream come true." Bruce Williams wanted a surrogate to bear his child, whom he would then give to Whitney to mother during the last few years of her life. "Why can't I make my own grand-child?" said Williams. "If your kid was dying and had a last wish, what's to say you wouldn't do something in the extreme to try to rectify the situation, especially since there isn't a great deal you can do against the disease itself?"

Anita Williams, Whitney's mother, hadn't known about her husband's ads in the magazine and strongly opposed his efforts to find a surrogate. "This is an eleven-year-old child who, like any eleven-year-old, is not ready for motherhood," she said.

According to the article, one woman—a twenty-seven-year-old resident of a Chicago suburb—had responded to Williams's ad. "He said he had no children, that he was a gay male, and wanted to father children," she said, on the condition of ano-nymity. "He would have visiting rights, which was fine with me. But he was very persistent about taking out large life insurance policies on himself, me, and the unborn baby. That scared me. It isn't something you say when you first meet somebody you want to have a child with."

On May 16, 1997, five years after Whitney Williams had been diagnosed with AIDS, I was saddened to learn that Whitney had died from her disease. Exactly how she had acquired HIV remains a mystery. Bruce Williams, his wife Anita, and their four children all tested negative for HIV.

Lesson: Be sympathetic, no matter how trying the circum-stance.

• • • •

On June 18, 2013, I appeared on *CBS This Morning* with Norah O'Donnell, Gayle King, and Charlie Rose to talk about megavi-tamins, dietary supplements, and alternative medicine. I had recently written a book titled *Do You Believe in Magic? The Sense and Nonsense of Alternative Medicine*, which warned about the dangers of the supplement industry. King was impres-sive. Typically, you're happy if the interviewer has read the title and flap of your book. King had read the entire book, which sat in front of her with paper tabs sticking out of the sections she had questions about. The conversation was amicable and informative. Then, suddenly, it wasn't.

In the book, I had written about Steve Jobs, who had died from pancreatic cancer two years earlier. Jobs didn't have the type of cancer for which no effective treatment exists (adenocarcinoma); rather, he had a neuroendocrine tumor—a cancer that can often be treated with early surgery. Unfortunately, instead of getting the surgery that might have saved his life, Jobs chose acupuncture, bowel cleansings, and fruit juices. When he finally had surgery, it was too late. Charlie Rose, as it turned out, was a friend of Steve Jobs. He was angered by my criticism of Jobs's choices, arguing that I wasn't Jobs's doctor, so how could I possibly comment on his care?

"With respect to you, sir," said Rose, "isn't it dangerous to say if you never treated a person what might or might not have been the consequence?"

"I'm sorry," I said. "I'm not understanding . . ." (Mild panic set in.)

"In other words, did you treat Jobs?" Rose insisted.

"No, I didn't," I said.

"So isn't it dangerous to suggest what he might have been able to do if you didn't treat him?" Rose asked.

I fell back on what I did know. "You know that neuroendocrine tumors have a 95 percent chance of survival with early surgery," I countered. "And you know that he significantly delayed his surgery. I think that adds up to the fact that he put himself at unnecessary risk by choosing an alternative course."

What I learned from this experience was not to go down the rabbit hole. The facts that I mentioned about Jobs's diagnosis and care weren't in dispute. They had been carefully described in Walter Isaacson's biography of Steve Jobs and in many news stories.

Lesson: Don't panic. The facts are your safety net.

. . . .

In January 2016 I was honored with the Franklin Founders Award for my work on the rotavirus vaccine. About two hundred people attended the event, which took place at a hotel in Philadelphia near Benjamin Franklin's grave. I was asked to give a twenty-minute talk, but I decided not to talk about the vaccine. (That was my first mistake.) Instead, I talked about an event that had occurred in Philadelphia twenty-five years earlier—one that I had written about in a book titled *Bad Faith: When Religious Belief Undermines Modern Medicine.*

In the first few months of 1991, Philadelphians suffered a massive measles epidemic. About fourteen hundred people were infected, and nine children died. The outbreak centered on two fundamentalist churches in the city: Faith Tabernacle and First Century Gospel—faith healing groups that didn't believe in vaccination or medical care. During the outbreak, children died from pneumonia because parents refused oxygen, or from dehydration because parents refused intravenous fluids. City public health officials reacted by passing a series of laws: first, allowing doctors to enter the homes of members of these fundamentalist communities to observe children from a distance; second, allowing doctors to physically examine children; third, requiring children to be hospitalized if necessary; and fourth, requiring children to be vaccinated. These laws were enforced even if the hospitalizations and vaccinations went against the parents' will.

It was tough time. When some of these children were admitted to our hospital, I had a chance to talk with the parents from these faith-healing groups. I asked them why they had chosen prayer over medicine. All parents answered the same way: "Jesus is our doctor," they said. No one likes to tell people how to raise their children. And surely no one likes to tell others how to practice their faith. But it seemed to me that allowing your child to die in the name of religion was a profoundly unreligious thing to do. During my talk, I said that public health officials in Philadelphia had been disappointed that none of the leaders in the Christian community had stood up and said, "This isn't us. Jesus loved children. He would never have wanted you to act this way." (The quote attributed to Jesus that

I cited during this speech was from Matthew 25:40: "Verily, I say unto you, inasmuch as you have done it unto one of the least of my brethren you have done it unto me." I argued that this quote could be emblazoned onto the entranceway of every children's hospital in the world.) I also specifically mentioned Anthony Bevilacqua, a Philadelphia-based bishop who at the time had been elevated to cardinal—the perfect person to step forward: a religious leader with a powerful voice.

Sitting in the front row of my talk was the priest (Father Tim) who had given the invocation for the ceremony, during which he had been amiable, generous, and pleasant, with a warm beatific smile. But he wasn't smiling anymore. When my talk was over, he was the first to raise his hand. He asked why I had targeted Christians when, in fact, all fundamentalist religions have a history of various kinds of abuse.

Got the picture. A priest wearing a collar—who only moments ago was the gentlest guy in the world—was now angrily confronting me in a room full of people. I explained that the reason that I had focused on Christian leaders was that both of these churches were acting in the name of Jesus. But it was uncomfortable, to say the least. One of the participants came up to me at the end and said, "You are one brave man." By which he meant that I was one stupid man. I probably should have stuck with a talk about the rotavirus vaccine. (I have since become friendly with Father Tim, who, as it turns out, lives only a block from me.)

Lesson: Take on religious issues at your own peril.

· · · ·

IN JULY 2013, THE PRODUCERS OF *INSIDE EDITION* ASKED ME TO talk about my book *Do You Believe in Magic?*, in which I warned against taking vitamins at doses vastly in excess of the recommended daily allowance. Called "megavitamins," these preparations can shift the body's balance between oxidation and anti-oxidation too far in favor of anti-oxidation. I talked about how oxidation was important in killing new cancer cells and preventing clogged coronary arteries. I talked about how people would have to eat fourteen oranges or eight cantaloupes to get the same amount of vitamin C as was contained in one 1,000 milligram tablet. I said that our stomachs are only so big for a reason and that we aren't meant to eat fourteen oranges or eight cantaloupes at once. I talked about the twenty studies that had consistently found that people who ingested large quantities of anti-oxidants, like vitamins A and E and the mineral selenium, increased their risk of cancer and heart disease and shortened their lives. Then I said that celebrities aren't always the best source of medical advice: "When Jenny McCarthy says don't get vaccines because that way you can avoid autism, that's terrible advice. When Suzanne Somers says use her hormone replacement therapy because it's all natural and isn't going to cause the sort of problems that regular hormone replacement therapy cause, that's bad advice. It's amazing to me that we turn to celebrities for health advice."

Later, the producers of *Inside Edition* asked Mehmet Oz to comment on my statements. Wisely, he refused. What would be the point? He had his own show and millions of devoted followers. I write books knowing that more people watch the Indianapolis 500 car race every year than read one book a year. Suzanne Somers, on the other hand, weighed in. Somers is probably best known for her role as Chrissy Snow on the television comedy *Three's Company*. She was also the spokesperson for the Thighmaster. At the time of the *Inside Edition* interview, Somers had become a one-woman industry, promoting megavitamins, supplements, and minerals, as well as skin-care, weight-loss, and detoxification products on her website. When Somers was asked to comment on statements that undermined her products, she said that I was just trying to build my reputation on her reputation. The night that the *Inside Edition* episode aired, I watched it with my wife and children. After Somers said that I was just trying to boost myself on her reputation, our children, who were twenty-one and eighteen years old at the time, looked at each other and said, "Who's Suzanne Somers?" The final episode of *Three's Company* had aired thirty years earlier.

Suzanne Somers isn't the only star with a Wild West medicine show. In 2008, Gwyneth Paltrow launched a lifestyle brand called Goop. Paltrow promotes vaginal steaming with mugwort to balance female hormones and cleanse the uterus. Apart from the fact that mugwort isn't a hormone and vaginal steam will never reach the uterus, vaginal steaming can cause burns and bacterial infections. Paltrow also promotes placing jade eggs the size of golf balls in the vagina overnight to boost orgasms, enhance kidney strength, and increase feminine energy, perhaps unaware that foreign bodies sitting in the vagina for long periods of time put users at risk of toxic shock syndrome. Finally, Paltrow promotes cleansing the colon with enemas to remove toxins and enhance the immune system. Colonic enemas, which have no proven benefit in otherwise healthy people, can cause dehydration, infections, vomiting, and, worst of all, bowel perforations.

Lesson: Avoid debating celebrities (unless no one remembers them anymore).

. . . .

WHEN YOU CHOOSE TO ENTER A CONTROVERSIAL ARENA, A BACKLASH is inevitable. After one television appearance, the CEO of my hospital received more than thirty phone calls asking him to fire me. I've had a man stalk me, after which my hospital hired lawyers to look into a restraining order. I've had my life threatened, following which my hospital investigated whether I was in any real danger. None of this could have been easy for the hospital administrators. Frankly, it would have been much easier for them to ask me to step back. But they didn't. The CEO at the time, Steven Altschuler, believed that I was doing the right thing— standing up for children in the same way that he was. Similarly, the hospital's next CEO, Madeline Bell, stood up for a policy requiring all health care workers to receive a yearly influenza vaccine as a condition of employment. This meant taking on the

Hospital Employees Union, but she did it. I consider myself lucky to be working at a place that has been willing to suffer the slings and arrows of controversy without flinching.

In 2002, just before the United States invaded Iraq, my hospital was tested again. At issue was the federal government's requirement to give all frontline medical personnel a smallpox vaccine.

Between 1998 and 2003, I served on the Advisory Committee on Immunization Practices (ACIP) at the CDC. The ACIP advises the CDC on how vaccines should be used. In 2002, months after the tragic events of September 11, 2001, the George W. Bush administration asked the ACIP to comment on whether a smallpox vaccine should be given to all health care workers. At the time, government officials were worried that rogue nations or groups in possession of smallpox would use it as a weapon. They wanted to make sure that Americans were protected. These discussions occurred one year before the United States invaded Iraq. (The smallpox vaccine had been discontinued as a routine vaccine for children in 1972 and as a routine vaccine for the military in 1982. Therefore, an entire generation of Americans was vulnerable to smallpox.)

Typically, the ACIP discusses many vaccine issues during each day of its two-day meeting. But in October 2002, we spent one full day talking about whether we should recommend the smallpox vaccine for frontline responders, like emergency room doctors and emergency transport personnel. When we finished, we put it to a vote. I thought that it didn't make sense to launch the program for several reasons. First, the smallpox vaccine, like the rabies vaccine, can protect people against disease even *after* they've been exposed. People will be protected against smallpox as long as they receive the vaccine within forty-eight hours of exposure. Second, smallpox isn't spread by small respiratory droplets—as measles is—it's spread by large droplets; so people would have to be face to face and within five feet of someone who is infected to catch it. Third, because smallpox always causes a rash, and because the smallpox rash invariably appears on the face, people would know that they had come into contact with someone who was infected. Finally, and most importantly, the smallpox vaccine is arguably our least safe vaccine. Side effects include encephalitis (inflammation of the brain), pericarditis (inflammation of the sac surrounding the heart), and overwhelming multisystem infection. About one of every million people inoculated with smallpox vaccine will die from the vaccination. My stance on the issue was as follows: Let's produce the vaccine; let's make sure that clinicians know how to give it; and let's alert frontline responders that they might need to get it. But let's wait until at least one case of smallpox appears somewhere on the face of the Earth before immunizing tens of thousands of people with a relatively dangerous vaccine.

When it came time to vote, the head of the ACIP, John Modlin, asked who was in favor of moving forward with the smallpox vaccination program. Eleven people raised their hands. Then he asked for all of those opposed. I raised my hand. When I looked up—I had been writing something down during both votes—I saw that I was the only one who had voted against the program. I had, inadvertently, put an "X" on my back. Both the media and the public attend ACIP meetings. Immediately after the session, representatives of the mainstream media crowded in on me, wanting to know why I had voted against the program. I ended up on several national television shows expressing my concern about the smallpox vaccine. On December 11, 2002, I appeared on a *60 Minutes* segment titled "The Most Dangerous Vaccine." Dan Rather was the host.

Dan Rather is an imposing figure, probably best remembered for his confrontation with Richard Nixon in 1974. Nixon was defending his policies only months before he would be impeached for his role in Watergate. Rather got up to ask a question. "Are you running for something, Dan?" asked Nixon. After much laughter and applause, Rather responded, "No, Mr. President.

Are you?" This was a time before journalists routinely challenged presidents to their face. Nonetheless, of all the journalists I have met during my dealings with the media, no one has been more generous, kind, or straightforward than Dan Rather. When Rather and his production team traveled down from New York City to my laboratory in Philadelphia, my children were off from school. So I brought them along, thinking how great it would be for them to meet such an iconic figure. Rather was wonderful with the kids, who were ten and eight years old at the time. He told them how proud they should be of their parents. He answered all of their questions about what a television journalist does. He couldn't have been more giving of his time.

My hospital's administrators were a little nervous about this interview. They were worried that I was about to go on a national television program and say that our hospital might ignore federal policy and withhold the smallpox vaccine from our frontline responders. They wanted to make sure that I stuck to the scientific issues and avoided talking about federal policy. So I did.

The head of my hospital's public relations team was Karen Muldoon Geus. Karen was a powerhouse. She had recently worked for the communications giant PrimeStar. But we had taken care of her baby, and she had fallen in love with the hospital. When we sat down for the interview, Karen positioned herself behind Dan Rather's right shoulder. That way, when I was talking to Rather, I could also see Karen out of the corner of my eye. When the interview started, Rather morphed into a tough interviewer. Every time he asked a question like "What would you advise George Bush?" or "If you were sitting in a room with George Bush, how would you change federal policy?" Karen would slowly shake her head from side to side. She wanted to make sure that I didn't include George W. Bush's name in my answers. So I didn't. I just stuck to the reasons that I thought mass smallpox vaccination didn't make sense. As the interview lengthened, the lights got hotter, and I got sweatier. I started to feel sorry for Richard Nixon. But I held on. In the end, I think Dan Rather got much of what he wanted, and I escaped without mentioning the name of the president.

Lesson: Make sure that you have the backing of your institution.

• • • •

UNLESS YOU HAVE A TEAM OF PUBLIC RELATIONS EXPERTS WORKING around the clock preparing you for every interview and press conference, you're going to say things that will be taken out of context—things that you wish you could take back. Even people who are supported by teams of public relations experts, like presidents and corporate executives, make this mistake.

The statement that I wish I could take back was my answer to the question, "How many vaccines could a child respond to at one time?" In the United States, children may receive as many as twenty-seven inoculations in the first few years of life and up to five shots at one visit. Many parents question whether babies can handle all of this. To answer this question, I asked an immunologist at the Wistar Institute in Philadelphia named Andy Caton to help me. He directed me to the work of two immunologists working at the University of California, San Diego: Mel Cohn and Rod Langman.

Cohn and Langman had determined the number of antibody-producing cells in the bloodstream, the number of distinct immunological components in viruses and bacteria, the quantity of antibodies that would be necessary to effectively respond to each immunological component, and the time it would take to do all of that. Using this method, they reasoned that the bigger the animal, the more antibody-producing cells in the blood and the greater the capacity to respond to a variety of immunological challenges. In other words, elephants can respond to more germs than hummingbirds. Using Cohn and Langman's model, I calculated that a child could probably respond to at least ten thousand vaccines at one time.

This number shouldn't have been surprising. The hepatitis B vaccine, for example, contains a single protein. The HPV vaccine is made using a single protein from each of nine different strains. So the HPV vaccine contains nine proteins. The measles vaccine contains ten proteins, the mumps vaccine nine, and the rubella vaccine five. So the MMR vaccine contains a total of twenty-four different proteins. And each of these proteins has distinct immunological regions called epitopes: usually, about ten epitopes per protein. The MMR vaccine, therefore, contains about two hundred forty different epitopes (twenty-four proteins times ten epitopes).

To put the challenge from vaccines in perspective, from the moment that babies leave the womb and enter the world, they have trillions of bacteria living on the surface of their nose, throat, skin, and intestines. Trillions. And each of these bacteria has between two thousand and six thousand proteins! And each of these proteins has about ten epitopes that could evoke an immune response. To keep these bacteria from invading our bodies and causing harm, we make about 5 grams of antibodies every day (which is a lot). Also, the food we eat, the water we drink, and the dust we inhale aren't sterile. So we are constantly bombarded with a wide variety of immunological challenges. Another thing: Common cold viruses, for example, reproduce themselves hundreds to thousands of times. Live, weakened viral vaccines, on the other hand, like the MMR, chickenpox, and rotavirus vaccines aren't well adapted to growing in our bodies. Instead of reproducing themselves thousands of times, these weakened viruses reproduce themselves much less efficiently. And most vaccines contain only parts of viruses or bacteria, so they don't reproduce themselves at all. Frankly, a scraped knee is a far greater immunological challenge than all of the childhood vaccines combined.

So when interviewers asked me how many vaccines a child could respond to at once, I said about ten thousand, which was actually a conservative estimate. Because we are making hundreds of millions of new antibody-producing cells every day, it probably would have been more accurate to say that we can respond to *at least* ten thousand.

Nonetheless, the anti-vaccine groups pounced. They interpreted the statement that children *could* receive the equivalent of ten thousand vaccines as meaning that they *should* receive that many. The image of children receiving ten thousand vaccines was hideous. I was a monster. People have challenged me to take ten thousand vaccines. One blogger said that my wife had given me ten thousand vaccines and that I had ended up in the emergency room. In July 2017, a PR news service declared that I had died from receiving ten thousand vaccines. (After the story was posted, a hospital representative called me to make sure I was still alive.) Indeed, when I was asked to speak to a group of journalism students at a local college, one of the students asked me what it felt like to get ten thousand vaccines. I assumed he was kidding. But he wasn't. He said that he'd read about it on the internet. I told him that people are allowed to lie on the internet. To this day, to some, I'm the children-should-get-ten-thousand-vaccines guy. There's no escape.

In retrospect, I probably should have just said that children's immune systems are vigorous and diverse, as they would have to be to navigate a world full of immunological challenges. I shouldn't have answered the question as an immunologist would have answered it, by assigning a number.

Lesson: You are going to say things that, although scientifically accurate, you will regret. It's unavoidable.

· · · ·

ALTHOUGH REPORTERS OFTEN TALK ABOUT HIDDEN AGENDAS AND conflicts of interest, few entities have a bigger conflict of interest than the media. Television news programs, especially, are under tremendous pressure to generate revenue. The logic is unassailable. Stories that are more interesting attract more viewers. More viewers mean higher ratings. Higher ratings mean more advertising dollars. And now, with news cycles that are twenty-four hours long, the television news beast must be fed constantly.

John Kerridge, a scientist searching for life on Mars, wrote, "If you want to be on television, tell them what you think they want to hear. If you want the public to know the truth, stick to print and radio." (The conception of newspapers, magazines, and television shows as voracious, insatiable, all-consuming beasts came from the pen of Evelyn Waugh in his 1938 book, *Scoop*, which featured a fictional newspaper called the *Daily Beast*.)

One particularly egregious feed-the-beast moment came on January 22, 1999, when ABC's *20/20* aired a segment that deeply scared the American public. The correspondent was Sylvia Chase. The subject was the hepatitis B vaccine. The teaser was ominous: "Next, an important medical controversy. Serious new questions about a vaccine most school children are forced to get: one given to millions of babies every year."

According to *20/20*, the hepatitis B vaccine was killing babies. "Thirty-three hours after his vaccination, thirteen-day-old Nicky Sexton's heart stopped," said Chase. "The coroner said it was sudden infant death syndrome, or SIDS. Lyla Belkin's death was also attributed to SIDS. She had received her first shot at six days old; the second, one month later." The camera slowly panned in on Lyla Belkin's gravestone: "Lyla Rose, Our Little Angel." The Belkin story was so heartfelt, so emotional, and so frightening that one of the nurses in my wife's pediatric practice came in the next morning and said, "That's it. I'm not giving the hepatitis B vaccine to babies anymore. I'm not going to participate in this charade."

Then Chase interviewed adults who believed that the hepatitis B vaccine had caused their multiple sclerosis, a life-long, debilitating illness. "Within three weeks of the third shot," said one woman, "I lost my vision." "I can't even feed myself," said another.

Neither of these concerns was remotely well founded. The hepatitis B vaccine was routinely recommended for all babies in the early 1990s. Throughout the next decade, as more and more babies received the vaccine, the incidence of SIDS didn't increase; it *decreased*, dramatically. The reason: the American Academy of Pediatrics' "Back to Sleep" program. Clinicians found that babies were much more likely to die from SIDS if they slept on their fronts instead of their backs. Indeed, if you compared the rate of hepatitis B immunization in the 1990s with the incidence of SIDS, you might conclude that the vaccine *prevented* SIDS. Also, studies had already been published showing that babies who had received the hepatitis B vaccine were not at greater risk of SIDS.

At the time that *20/20* aired its report, Barbara Walters and Hugh Downs hosted the show. Walters ended the segment with a comment. "What a choice for parents to have to make," she declared, incredulous. "There is so much conflicting information." The problem for parents, however, wasn't that there was so much conflicting information; it was that there was so much misleading information.

After the program aired, two studies in the prestigious *New England Journal of Medicine* showed that the hepatitis B vaccine neither caused nor worsened the symptoms of multiple sclerosis. A few months later, I ran into a junior producer from ABC at a conference. I asked whether her network would be willing to do a follow-up program explaining that the hepatitis B vaccine didn't cause multiple sclerosis. She told me that they wouldn't be interested, but thanks. I said that I thought that it was her responsibility to reassure parents, given that her show had just scared them unnecessarily. The discussion got more animated. Loud, even. Clearly, from the standpoint of the producer, I wasn't getting it. I was either too naïve or too uninformed to understand how television magazine programs really worked. Finally, completely exasperated, she said, "Look, our job is to be interesting. If it also happens to be true, great!"

It seems to me that the junior producer's statement can be interpreted in three ways. First, the cynical way. The fact that the hepatitis B vaccine was safe and effective wasn't particularly interesting; the possibility that the vaccine was instantly killing children or permanently disabling adults was. Right or wrong, the idea was interesting. And, right or wrong, it would sell advertising. If children suffered because of bad reporting, so be it.

Second, the more generous way. The producer felt sympathy for parents dealing with the death of a child and for young adults confronting a life-long illness. The *20/20* program gave these people a chance to air their concerns. And if those concerns turned out to be true, then the producer had done a service. If they turned out to be wrong, at least the producer had tried.

Third, the dispiriting way. The producer lacked a sufficient grasp of the difference between problems causally or coincidentally associated with a vaccine. If one event preceded another, then it must have caused it. The rooster crows; the sun comes up. The rooster crows again; the sun comes up again. Therefore, the rooster is causing the sun to come up. A few years later, Katie Couric showed that this last interpretation was probably the right one.

Katie Couric worked for NBC News from 1989 to 2006, for CBS News from 2006 to 2011, and for ABC News from 2011 to 2014. She was the co-host of *Today*, an anchor for *CBS Evening News*, and a correspondent for *60 Minutes*. She also hosted *Katie*, a syndicated daytime talk show, from 2012 to 2014. Katie Couric was a trusted, veteran newscaster. But when she took on the HPV vaccine on *Katie*, she seriously damaged her reputation. The mea culpa that followed provided an interesting (and frightening) insight into how at least some television journalists think.

At the time of licensure, the HPV vaccine had been tested in about thirty thousand people over seven years. It was safe and effective. Since licensure, the vaccine has been formally tested for safety in about a million people. No other vaccine has been tested more extensively. The only side effect that has been consistently associated with HPV vaccine is fainting. And teenagers don't even have to receive the vaccine to faint; sometimes just unsheathing the needle does it, or being next in line.

Between 1998 and 2003, I was a member of a CDC advisory committee that discussed whether the HPV vaccine should be

recommended for all children. The question before the committee was when would be the best time to give it. Because the vaccine prevents but doesn't treat HPV, it has to be given *before* the
first sexual encounter. Epidemiologists at the CDC presented
studies showing that about 25 percent of girls had what they
(non-euphemistically) called "penetrating sex" by the time they
were fifteen years old. The term they used was "sexual debut,"
making sex sound like a public performance. (Thank you.
You've been a wonderful audience. I'll be here all week.) The
sexual-debut data weren't limited to one particular group; the
25 percent estimate applied to those living in urban, suburban,
and rural settings and to those of Latino, Caucasian, Asian-
American, and African American ethnicity—in other words, to
everybody. For these reasons, the CDC recommended that all
adolescents receive the HPV vaccine between eleven and thirteen years of age.

On December 4, 2013, Katie Couric aired an episode that
raised serious questions about the HPV vaccine. The show
featured two women whose children had supposedly been
harmed by the vaccine. In one case, the child had died following
vaccination. In the other, the child had suffered chronic fatigue
and was unable to get out of bed to go to school. Both mothers
were sure that the HPV vaccine had been the cause. Couric
didn't mention that several studies had already shown that the
incidence of teenage death and the incidence of chronic fatigue
symptoms were no different in adolescents who had received the
vaccine compared with those who hadn't. She also failed to
mention that the girl who had died suddenly had twice been
found to have an electrical conduction abnormality in her heart:
a known cause of sudden death in adolescents and young adults.
And she neglected to mention that this same girl had complained
of feeling dizzy and faint several months *before* she had received
her third dose of the HPV vaccine: the dose that her mother
claimed had killed her. When Couric asked whether the girl had
suffered any symptoms before she died, the mother said that she
wasn't allowed to answer because she had a "case pending."

Nonetheless, information about this girl's heart problems was available at the time of the interview. It was disappointing to see such poor journalism from such a trusted journalist.

Angry that she had chosen to scare parents away from a life-saving vaccine, the medical and scientific community rained down on Katie Couric. In response, Couric wrote what she believed was an apology, but it didn't sound like much of one: "Following the show, and in fact even before it aired, there was criticism that the program was too anti-vaccine and anti-science, and in retrospect, some of that criticism was valid." (This was the apology part.) "We simply spent too much time on the serious adverse events that have been reported in very rare cases following the vaccine." (This was the lame-apology part. Couric wasn't putting forward a strong statement that the vaccine didn't cause terrible side effects. She was just saying that she'd spent too much time talking about the awful side effects, one of which was death.)

Although Katie Couric ultimately concluded that, "based on the science, my personal view is that the benefits of the HPV vaccine far outweigh its risks," her statement continued to defend a national television broadcast that was indefensible: "Concerns have been raised about reactions to the vaccination. Unfortunately, there's no question reactions can occur, as with all vaccines. . . . As a journalist, I felt that we simply could not ignore these reports." She never helped her viewers think through the difference between causal and coincidental associations. Katie Couric's program dramatized the fact that the HPV vaccine doesn't prevent fatal heart arrhythmias or chronic fatigue syndrome. It only prevents HPV infections.

At the end of her lame apology, Couric further confused her followers: "Our goal in doing this show was to help parents make an informed choice about the HPV vaccine, not cause irrational fear." But providing misleading and frightening information can *only* cause irrational fear. Perhaps the most disappointing part of Couric's statement was that she has been a zealous advocate for preventing colon cancer, her husband having died from the disease.

Although, to my mind, *20/20* and *Katie* would both be in the running for the coveted titled of "Most Misleading Science News Program," the award must go to CNN for its coverage of a "miracle cure" for AIDS. The story aired during sweeps week, when networks set their prices for advertising.

On June 2, 1990, CNN's *Health Week* breathlessly told the story of two doctors who had supposedly found a cure for AIDS. Both worked at a small hospital in Atlanta. Their first patient, Carl Crawford, had been suffering from an AIDS-related skin cancer called Kaposi's sarcoma. To "cure" his AIDS, the doctors put Crawford under general anesthesia, periodically removed several pints of his blood, heated the blood to 108°F, returned it back to his body, and watched as his temperature rose to dangerously high levels. (Extremely high fevers can cause brain damage.) After two hours, Crawford was cooled down. When the procedure was over, the doctors claimed that HIV had been eliminated from Carl Crawford's body—an unprecedented scientific and medical achievement. Also, Crawford's skin cancer had begun to disappear: remarkable, given that no known treatment for Kaposi's sarcoma existed. The story sparked a media firestorm; hundreds of television, radio, and magazine outlets retold the tale. Soon the Atlanta doctors were receiving more than a thousand requests a day from patients desperate for a cure. The story was so universally embraced that the media covered the heat treatment of a second AIDS patient live on national television.

Of interest, this wasn't the first time that heat had been used to treat a devastating infection. In the early 1900s, the "fever box" successfully treated syphilis. (This therapy, which won a Nobel Prize for its inventor, was featured on the short-lived Cinemax series *The Knick*, starring Clive Owen.) In the days before antibiotics, fever therapy was also used to treat arthritis caused by gonorrhea. Before 1990, however, fever therapy had never been used to treat viral infections like HIV. There was a reason for this. Viruses, unlike the bacteria that cause syphilis and gonorrhea, reproduce *inside* cells. So it's virtually impossible to kill a virus with heat without first killing the patient's cells. Or, said another way, without first killing the patient.

It didn't take long for the Atlanta miracle to fall apart. On August 3, 1990, a third patient to undergo heat treatment died following the procedure. The National Institutes of Health sent a team to investigate, which found no immunological, virological, or clinical evidence that heat therapy cured AIDS. Also, Carl Crawford's skin cancer turned out not to be skin cancer. When the dust settled, one of the Atlanta doctors lost his license to practice medicine, and the small Atlanta hospital with the miracle AIDS cure closed its doors.

Lesson: Advertising revenue is the ultimate conflict of interest.

CHAPTER 5

To Debate or Not to Debate

You can sway a thousand men by appealing to their prejudices quicker than you can convince one man by logic.

—ROBERT A. HEINLEIN, *REVOLT IN 2100/METHUSELAH'S CHILDREN*

Launched in 1982, the political talk show *Crossfire* aired on CNN for almost twenty-five years. The format was simple. Political pundits on the left squared off against their counterparts on the right. The show was lively and entertaining. At the heart of these political debates was whether Americans benefited from more or less government. The show was successful in large part because of its format—television thrives on controversies, especially ones that will never be resolved.

Science, on the other hand, isn't politics. Once scientific truths have emerged, they aren't debatable. Nonetheless, scientists are occasionally asked to debate them. In this chapter, using vaccines, evolution, homeopathy, and the Holocaust as examples, I'll discuss whether these debates are worthwhile or whether they just provide another venue for people to express their ill-founded, disproven—and, in the case of Holocaust denialism—heinous beliefs.

. . . .

ON FEBRUARY 5, 2015, I WAS ASKED TO APPEAR ON A TELEVISION
program called *Democracy Now!* hosted by Amy Goodman.
Appearing with me were Dorit Rubinstein Reiss, a law professor
at the University of California Hastings College of the Law,
and Mary Holland, a lawyer who contributes to the anti-vaccine
blog *Age of Autism* and a co-editor of the book *Vaccine
Epidemic: How Corporate Greed, Biased Science, and Coercive
Government Threaten Our Human Rights, Our Health, and
Our Children.* Needless to say, Mary Holland is not a big fan
of vaccines.

Dorit Reiss was the first to be interviewed. She talked about
how the legal system has tried to balance a parent's right to act
in the best interest of their child with the child's right to a long
life. She carefully and thoughtfully explained how states have
worked to protect their citizens from outbreaks of preventable
infections while at the same time allowing religious and philosophical exemptions to vaccination. I learned a lot from her.

Next up was Mary Holland, who proceeded to pepper the
viewing audience with one misleading comment after another.
Holland said that "vaccines [cause] severe injury, brain damage
in particular." (Vaccines don't cause permanent brain damage.
Some of the diseases that vaccines prevent can cause permanent
brain damage.) Regarding vaccine makers, Holland said that
"we now have an industry with high profits and exceedingly low
litigation risks." (If vaccines are so profitable, why have companies largely abandoned them? Today, only four companies
make vaccines for America's children. In 1955, twenty-seven
companies made them. While mergers account for some of this
attrition, most was caused by dropout.) Then Holland taught
the viewing audience that "there's no question that [vaccines]
can compromise the immune system." (Vaccines strengthen the
immune system.) Holland said that the "MMR vaccine [causes]
precipitous [developmental] regression." (Seventeen studies have
shown that it doesn't.) Holland warned of a conspiracy by
the federal government to hide the truth about vaccines:
"Dr. William Thompson, in the CDC, has come forward and

said that he colluded with other key scientists to mask a signal that vaccines and autism are linked." (No, he hadn't.) Then we learned that "boys, neonates, who received the hep. B vaccine were nine times more likely to end up in special education [and] three times more likely to have an autism diagnosis." (And so the goalpost shifts. Now it's not the MMR vaccine that causes autism; it's the hepatitis B vaccine. Also untrue.) Holland then stated, "I'm not anti-vaccine." (If you vilify vaccines with little regard for scientific accuracy, it's fair to say you're anti-vaccine.) We learned that "children are too young to be vaccinated because the risks of injury are so great." (Young children are vaccinated because many of the diseases vaccines prevent occur in young children.) Next, we learned that "we don't have herd immunity for measles. We don't have herd immunity for any of these diseases." (Measles was eliminated from the United States in 2000 even though less than 100 percent of the population had been vaccinated. That's what herd immunity is.) Holland said, "We have not looked, either retrospectively or prospectively, at what is the health of children who have gotten the CDC-recommended schedule, what is the health of children who have gotten the state mandate, and what is the status of children who are unvaccinated. That's a totally doable study." (Two studies have examined children who either received vaccines according to the CDC schedule or whose parents had chosen to delay vaccines. Developmental outcomes were the same in both groups.) Holland then said, "The way that vaccines are tested, before they're recommended and then mandated, is they're tested individually. They're not tested as part of a schedule." (Before any new vaccine is licensed, the FDA requires manufacturers to prove that the vaccine doesn't interfere with the safety or immunogenicity of existing vaccines and vice versa.) Finally, we learned that the "hepatitis B [virus] is sexually transmitted. What is the rationality of giving a baby a vaccine that will wear off by the time they're sexually active?" (Before the hepatitis B vaccine was routinely given to newborns in 1991, every year about eight thousand children less than ten years of age contracted the infection from casual contact with relatives who didn't know that they were infected. Also, the immunity induced by the hepatitis B vaccine lasts at least thirty years and probably a lifetime.)

It was hard for me to watch Amy Goodman and Mary Holland talking to each other in the *Democracy Now!* studio while I was sitting in Philadelphia, connected via remote video link. While waiting, I wrote down each of Holland's misleading statements. When Amy Goodman welcomed me to the show, I started by saying, "I honestly think the last ten minutes of your program set a new record for consecutive statements that were incorrect." (This is called endearing yourself to the host.)

To her credit, Goodman allowed me to address many of Holland's statements. But when I took exception to the format of the show, she pushed back. "I mean, to be fair," said Goodman, "we wanted to have both of you on together to have a conversation, because there are many in this country, and a growing movement of parents, who are deeply concerned. But you wanted to have this conversation separately—Mary Holland and you separately. So, it's important, I think, [to] have this kind of dialogue on all of these issues."

Again, Goodman gave me a chance to respond, so I gave her my opinion: "I think that it is not important to have a debate about the science with someone who clearly doesn't know the science. I'm sorry. Ms. Holland misrepresented the science again and again and again. I don't think that in any way helps your viewers. I don't think it's fair to have a debate where two sides are represented, when only one side is supported by the science. I'd like to think we're beyond that." I don't know why I said that I thought that we're beyond that. We're not beyond that at all. The media loves controversy. Thrives on it. And while it's fair to have debates about religion or philosophy or politics in which two sides of an issue can reasonably be debated, it's not fair to the viewers to debate scientific issues when the science is settled. When you agree to do that, you're agreeing, at least tacitly, that the issue *is* debatable. Five minutes into the program, no one remembers who the scientific expert is. And they remember the fight far more than the facts. Good for television; bad for science—and very bad for the viewers.

The three experts on *Democracy Now!* played different roles during this debate. Professor Reiss spoke on her training: the law. Mary Holland, another legal expert, followed Reiss. But instead of Holland challenging Reiss on the law, she talked about science, which wasn't her training. By placing me in the position of responding to Holland, the show elevated Holland from someone with strong opinions on vaccines to someone with medical expertise in vaccines.

So my recommendation is to avoid these situations. But as you'll see in the following stories, other scientists have successfully taken on science denialists while at the same time educating their listeners. So I think that my advice here might be wrong.

On February 5, 2014, Bill Nye ("the Science Guy"), a mechanical engineer and science educator, debated Ken Ham, the CEO of Answers in Genesis: a fundamentalist Christian organization that advocates for a literal interpretation of the book of Genesis. Many scientists have argued that Nye should never have agreed to do this. "Inevitably, when you turn down the invitation, you will be accused of cowardice or of an inability to defend your own beliefs," wrote Richard Dawkins, an evolutionary biologist and popular author. "But that is better than supplying the creationists with what they crave: the oxygen of respectability in the world of science."

The debate, titled "Is Creation a Viable Model of Origins?," took place at the Creation Museum in Petersburg, Kentucky, which is operated by Answers in Genesis. This museum features, among other impossibilities, a diorama of a dinosaur standing next to a small child. Dinosaurs became extinct about sixty-five million years ago. *Homo sapiens* (us) have been around for only about two hundred fifty thousand years. So dinosaurs and people were never alive together on this planet.

Nine hundred people attended the debate, three million viewed it live online, and ten million have since seen it on YouTube.

Ken Ham opened the discussion by claiming that everything we need to know about the creation of the Earth and humans can be found in Genesis, a book written more than two thousand years ago. Given what we know from Genesis, claimed Ham, Charles Darwin was wrong. Humans and all manner of life on the planet didn't evolve from a common ancestor. Rather, God created humans and animals much as we see them today. Nye countered that to deny evolution meant that you had to ignore about two hundred fifty thousand years of fossil records. Ham countered that Nye couldn't possibly know that because he wasn't there to observe it, implying that the fossil records were probably fakes.

Ken Ham then claimed that the Earth was only six thousand years old: "From Adam to Abraham—you've got two thousand years; from Abraham to Christ, two thousand years; from Christ to the present, two thousand years. That's how we reach six thousand years."

Nye countered that carbon dating of rocks showed that Ham was off by about 4.5 billion years.

Ham countered that carbon dating methods were unreliable. "The only infallible dating method comes from the only witness who was there: God," said Ham. "And his word is all that is reliable."

There was, of course, no way that Bill Nye could have reasonably argued the existence of God. So he didn't. Rather, he countered with several interesting facts. Nye said that ice cores in places like the North Pole show six hundred eighty thousand layers—which would have taken at least six hundred eighty thousand cycles of winter and summer to create. If Ham was correct in stating that the Earth was only a few thousand years old, then hundreds of winter–summer cycles would have had to have occurred every year, something that Nye thought people would have noticed. "If we accept that the Bible as translated into English serves as a science text," said Nye, "it means that

Mr. Ham's interpretation of those words, which have been retranslated again and again over three millennia, is somehow to be more respected than what you can observe in nature, what you can observe literally in your own backyard here in Kentucky. It's a troubling and unsettling point of view."

Nye also took on Ham's belief in the Great Flood and Noah's Ark. He asked the audience whether they found it credible that eight men without any previous ship-building experience could have built a wooden boat five hundred feet long, knowing that no wooden boat that long has ever been built. And knowing that a wooden boat built to the parameters described in the Bible could never float. Indeed, the longest wooden boat ever built was about three hundred feet long, a six-masted schooner called the Wyoming, which sank soon after it was put to sea. Further, Noah's Ark supposedly contained eight zookeepers and fourteen thousand animals. According to Ham, those fourteen thousand animals, paired male and female, represented seven thousand species. Given that there are sixteen million species of animals on the Earth today, Nye said this would mean that eleven new species would have had to have been created every day, when in fact species are being eliminated, not created.

Ham countered that Nye "can't say that Noah couldn't build the Ark because you never met Noah." Again, Ham argued that Nye couldn't comment on the past because he wasn't there to observe it.

Continuing his attack on the credibility of the Great Flood, which supposedly occurred about four thousand years ago, Nye showed a series of trees currently on Earth that were more than six thousand years old, including one in Sweden that was more than nine thousand years old—impossible given the Great Flood's timeline.

Nye also said that millions and millions of Christians alive today believe in evolution, which didn't make them bad Christians. Ham countered that Nye, by supporting the fact of evolution, was committing an act against God. That God, not humanity, was the ultimate authority.

By teaching evolution, Ham said that Nye was "imposing an anti-God religion on a generation of unsuspecting students."

The debate lasted for two and a half hours. Throughout the event, Nye was calm and respectful. At the end, however, his anger flashed, albeit briefly. "I just want to close by reminding everybody what's at stake here," he said. "If we abandon all that we've learned, our ancestors, what they've learned about nature and our place in it, if we abandon the process, if we let go of everything that we've learned before us, if we stop driving forward, stop looking for the next answer to the next question, we in the United States will be outcompeted by other countries, other economies. Now that would be OK, I guess. But I was born here. I'm a patriot. And so we have to embrace science education. To the voters and taxpayers who are watching, please keep that in mind. We have to keep science education in science classes. You don't want to raise a generation of science students who don't understand natural laws."

During the question-and-answer period, an audience member asked Nye about the Big Bang theory. Nye explained that radio waves in space contain a static noise consistent with an earlier explosion and that the galaxies are constantly moving away from us in all directions. This, explained Nye, was the origin of the universe. The questioner wanted to know about the source of matter *before* the Big Bang. Nye loved the question. Excitedly, he said, "We don't know!" But Nye reveled in the fact that the questioner wanted to know more. This, to Bill Nye, was the essence of science: curiosity leading to questions leading to new knowledge.

Ken Ham ended Nye's excitement with a dull thud. "There's a book out there that answers this question," said Ham. "And it's in the first sentence: 'In the beginning, God created the heavens and the Earth.' "

"God," concluded Ham, "was necessary for science."

On November 26, 2012, Joe Schwarcz debated André Saine in front of a general audience at McGill University in Montreal. The title of the debate was "Homeopathy: Mere Placebo or

Great Medicine?" Schwarcz, who earned his PhD in chemistry in the early 1970s, is now a professor of chemistry at McGill, as well as the director of McGill's Office for Science and Society. The author of twelve books written for the general public about chemistry in daily life, Schwarcz has been a wonderful advocate for science. André Saine, a homeopath, was the president of the Quebec Society of Naturopathic Medicine. For more than twenty-five years, Saine had made it his mission to educate the press and the public about the wonders of homeopathy.

Early in the debate, we learned that homeopathy was the brainchild of Samuel Hahnemann, a German physician who had wearied of the cruel and destructive medical procedures of his day, like leeches, bloodletting, and scarification. In the late 1700s, Hahnemann had an epiphany. While chewing on a piece of cinchona bark, he developed a high fever. Hahnemann knew that cinchona bark was a treatment for malaria (because it contains quinine), and he also knew that high fevers were a symptom of malaria. So he invented the first pillar of homeopathy: *like cures like.* In other words, patients should be given emetics to cure vomiting and cathartics to cure diarrhea and pyretics to treat fever. (*Homeopathy* literally means "similar suffering.") The obvious problem with this idea is that giving a medicine that causes vomiting to patients who are vomiting will only make them worse. So Hahnemann invented the second pillar of homeopathy: *the law of infinitesimals.* Instead of giving the patient an actual drug, Hahnemann diluted the drug in water or alcohol over and over again, shaking the vial vigorously with each dilution. When he was finished, not a single molecule of the original drug remained.

Joe Schwarcz mentioned one example of a modern-day homeopathic remedy called Oscillococcinum. Billed as "Nature's #1 Flu Medicine," Oscillococcinum is made by taking the liver and spleen of a duck (*Anas barbariae*), homogenizing it, diluting it to a ratio of 1:100 in water, then repeating the 1:100 dilution two hundred more times. The final dilution would then be 10^{-400}. In other words, the duck is gone.

Modern-day homeopaths understand that their highly diluted preparations don't contain the active ingredient. They argue, however, that the water in which the product was diluted *remembers* that the medicine had been there. (The Earth contains a limited amount of water. You don't want it to remember where it's been.)

André Saine opened the debate with a simple declaration: "Homeopathy works!" Saine argued that it works on cells and on plants and on animals and on people. He said that homeopathic remedies could treat attention deficit disorder or serve as adjunctive therapy for bacterial sepsis. He said that Joe Schwarcz was probably going to get up after him and tell the audience that randomized, placebo-controlled trials of homeopathy have shown that it doesn't work any better than a placebo. But the problem with those trials, according to Saine, was that the homeopathic dosing in those studies hadn't been individualized for each patient. Only homeopaths know exactly how much of each substance to give. And the studies showing that homeopathic remedies weren't any better than a placebo hadn't been performed by homeopaths; they'd been performed by conventional practitioners who obviously didn't know what they were doing. Saine said that Samuel Hahnemann had been a rigorous, thoughtful, hard-working scientist who was admired in his time and should still be admired today. He said that although he agreed that homeopathic substances are diluted to the point that the active ingredient is gone, something strange happens—something he couldn't quite explain. Something mysterious and otherworldly. Somehow the original substance still exerts an influence. Even more interesting, this effect appears to enhance over time. A miracle.

Joe Schwarcz sat patiently listening to Saine's argument. When it was his turn to speak, he commented on the title of the program. "We've asked the wrong question," he said. "The title 'Homeopathy: Mere Placebo or Great Medicine?' implies that placebos *can't* be an effective medicine. In fact, they can be both." Schwarcz explained the power of placebos, alluding to studies showing that people, thinking they're getting a real medicine, can learn to release their body's own pain-relieving chemicals (called endorphins), or gamma-interferon (which stimulates the immune system), or cortisol (which suppresses the immune system), or dopamine (which affects brain function). Schwarcz agreed that placebo medicine had a place. "There's nothing wrong with having an effective placebo," he said, "as long as it is honestly delivered and as long as the patient isn't being tricked that something else is happening."

Schwarcz made a number of other interesting points. He said that water molecules do indeed change in the presence of an active substance—but that this change lasts for only about one picosecond (one trillionth of a second). And he disparaged studies showing that homeopathic remedies could treat attention deficit disorder. Schwarcz said that science stood on several pillars: peer review, critical thinking, reproducibility, and plausibility. The weakest pillar was peer review. Schwarcz said that six thousand five hundred medical publications produced an average of four thousand papers a day! Not surprisingly, some papers are excellent, some are awful, and most are mediocre. In other words, it wasn't hard to find some journal somewhere in the world that would publish a study supporting anything, no matter how inane or implausible. Saine, argued Schwarcz, was relying on those poorly constructed studies to support his claims. "I admit that these are anomalous findings," said Schwarcz. "But that's not the point. The point is, so what?"

Then Schwarcz took homeopathy to its illogical end, referring to a homeopathic product made from fragments of concrete obtained from the Berlin Wall to reduce anxiety. Homeopaths who used this product had reasoned that the Berlin Wall had caused much anxiety. Therefore, according to Hahnemann's theories, dilute preparations of the Wall should reduce anxiety. Schwarcz said that he assumed that Saine didn't buy into this particular hoax. Or into homeopathic products that claimed to "cure" people who were gay.

Only once during the debate did Schwarcz lose his temper. It was when he criticized homeopaths for claiming that their products cured malaria, AIDS, and cancer and that they prevented infections like HPV (homeopathic vaccines are called nosodes). Schwarcz argued that practitioners who offered homeopathic remedies for diseases for which actual treatments existed shouldn't be tolerated. Rather, these homeopaths should be held accountable for profiting from a medical fraud. Schwarcz then told the story of an Australian woman with a treatable form of colon cancer who had died because she had chosen homeopathic remedies instead of lifesaving chemotherapy.

During the rebuttal stage of the debate, Saine became more animated, insistent. "Science is passing you by, Joe," he said. "Serious scientists are supporting our point of view. In the future, all of us will be practicing homeopathy."

Schwarcz countered by noting that throughout history, every new therapy has been met with skepticism. He argued that initially no one believed that vitamin C could cure scurvy, or that vitamin D could cure rickets, or that aspirin could reduce inflammation, or that the foxglove plant, which contains digitalis, could treat heart failure. But clinicians don't debate these issues anymore because "the evidence has become overwhelming." "Almost every new medicine in history has been opposed," said Schwarcz. "New technology breeds criticism. But eventually the truth comes out. So why are we still arguing about something that we were arguing about two hundred and fifty years ago?"

Because it is biologically impossible, homeopathy is often the subject of parody. For example, a group called the Centre for Inquiry Canada created an ad promoting its anti-homeopathy stance, which showed two parents sitting among a group of eight of their children. "Homeopathic contraception," the ad reads, "There's still nothing in it."

Comedians have also weighed in. In the United Kingdom, the comedians David Mitchell and Robert Webb featured a segment titled "Homeopathic A&E" on their sketch show, *That Mitchell and Webb Look* ("A&E" referring to a hospital's accident and emergency department). The sketch opens with a man severely injured in a car accident being wheeled into the "Homeopathic A&E" on a gurney. The doctor and nurse, seeing the severity of the man's injuries, are frantic. "Give me a solution of arnica . . . one part in a million," shouts the doctor.

"Are you sure?" asks the nurse. "It looks serious."

"You're right," the doctor responds. "We need to strengthen the dose. One part in ten million."

"Does anybody know what sort of car hit him?" the doctor implores. Once informed, he instructs the nurse, "Get me a bit of blue Ford Mondeo. Put it in water, shake it, dilute it, shake it again, dilute it again, do some more shaking, dilute it some more, then put three drops on his tongue." He concludes, "If that doesn't cure him, I don't know what will."

Despite their efforts, the patient sits up, looks quizzically at the doctor, and dies. "Time of death, 3:34—ish," says the doctor, noting the sundial on the wall.

In the next scene, the doctor is sitting at a bar, comforting himself with a homeopathic beer (one drop of beer in a large stein of water). "Sometimes I think a trace solution of deadly nightshade or a statistically negligible quantity of arsenic just isn't enough," he laments to a colleague.

"That's crazy talk, Simon," says his friend. "OK, so you kill the odd patient with cancer or heart disease or bronchitis, flu, chicken pox, or measles. But when someone comes in with a vague sense of unease or a touch of the nerves or even just more money than sense, you'll be there for them—with a bottle of basically just water in one hand and a huge invoice in the other."

On November 17, 2013, Michael Shermer debated Mark Weber at the Institute for Historical Review in Newport Beach, California. At issue was whether the Holocaust actually happened. Twenty years earlier, Shermer had engaged in a similar debate on the *Phil Donahue Show*. Donahue was the first mainstream talk show host to give Holocaust deniers a voice. Donahue did it because, as he claimed on his show, a recent poll had shown that 22 percent of Americans believed it was possible that the Holocaust had never happened, and 12 percent said that they didn't know. Although many of his fellow historians disagreed with him, Shermer argued that it was important to debate the issue of Holocaust denialism in public. "Not only is it defensible to respond to the Holocaust deniers," he wrote, "it is, we believe, our duty. The Holocaust deniers have succeeded in spreading their beliefs in the media and in the academic world. They are featured on national and local TV and radio talk shows, are invited to speak on college campuses, and have succeeded in placing full-page paid advertisements in college and university newspapers, including those of Brandeis University, Pennsylvania State University, and Queens College. Some of these ads arguing that the Holocaust never happened ran without comment."

Michael Shermer is a professor of history at Occidental College in Los Angeles, the founder of the Skeptics Society, and the editor-in-chief of *Skeptic* magazine, a publication that promotes scientific literacy. Shermer is also the author of several books, the most popular of which is *Why People Believe Weird Things*.

Mark Weber was an undergraduate at the University of Illinois in Chicago. He subsequently trained at the University of Munich, then Portland State University, where he received a bachelor's degree in history, and finally Indiana University Bloomington, where he received his master's degree. Weber has appeared on talk shows ranging from the *Montel Williams Show* to *60 Minutes*. Currently, Weber heads the Institute for Historical Review. Founded in 1978 with a grant from Thomas Edison's grandniece, Jean Edison-Farrel, the Institute is considered to be the international center for Holocaust denialism. Weber, the editor-in-chief of the Institute's journal, the *Journal of Historical Review*, took over as director of the Institute in 1995.

The setting for the debate was the Institute's annual conference, usually attended by about two hundred fifty people. Typically, the display table features books like Adolf Hitler's

Mein Kampf, The Protocols of the Elders of Zion (an essay claiming that Jewish elders had conspired to gain control of the world by fixing the price of gold), and Henry Ford's anti-Semitic screed, *The International Jew.*

Greg Raven, the associate editor of the *Journal of Historical Review*, introduced the debate saying, "The Holocaust was one of the most emotion-laden and propagandized chapters in contemporary history." He then offered Shermer a chance to call the coin flip to determine who would speak first. Shermer won, but elected to go second. Mark Weber began by laying out his three basic claims. First, no German policy or program ever existed to exterminate Europe's Jews. Second, stories of mass killings and gas chambers are mythical; gas chambers were used for delousing clothing and blankets, not for killing Jews. Third, six million deaths is a vast over-exaggeration; only between three hundred thousand and two million Jews died in the camps.

While Weber agreed that Jews had been persecuted under the Nazi regime, he argued that the goal was to deport Jews and have them work in labor camps, not to kill them. Although Weber didn't deny that many Jews died in those camps, he said that their deaths were not caused by state-sponsored killings; rather, they were caused by overwork, starvation, and diseases like typhus. Indeed, according to Weber, the Nazi commandants who ran the camps and the Nazi doctors charged with maintaining the health of their inmates were chastised by the German high command for not doing a better job of keeping inmates alive. Yes, the Jews had committed crimes against the Nazi regime. And yes, the Jews had been punished for those crimes by deportation. But Germans weren't interested in killing Jews. Why would they be? The Germans needed the Jews as laborers to help with the war effort. Weber argued that the term *Final Solution* referred to mass deportations, not mass exterminations.

Weber went on to say that all of the statements made in books, films, and personal testimonials by survivors of Auschwitz were lies put forward by a propaganda machine run by the Anti-Defamation League and other Jewish groups. Indeed, experts who have examined the remains of Auschwitz have stated that no gas chambers existed. Weber concluded, "The Holocaust is a flourishing business and even a kind of new religion for many Jews," calling the unfounded phenomenon "Holocaust-mania."

Then it was Michael Shermer's turn. Shermer listed evidence for the systematic, state-sponsored genocide of Europe's Jews. Although he was calm and reasoned, Shermer was clearly nervous, beginning his rebuttal by stating that he wasn't Jewish. He also said that despite claims to the contrary, the Anti-Defamation League didn't support his magazine, *Skeptic*. Shermer said that the Holocaust "was a known crime that no one doubts, except for you guys."

Shermer read quotes from Hans Frank, the Nazi official principally responsible for instituting the reign of terror against Poland's Jews. Frank had lamented to German officials that he "couldn't kill all of Poland's Jews in one year, but if you help me this end can be obtained." Later Frank said, "We must annihilate the Jews, so rid yourself of pity." Shermer argued that if the goal was merely to deport Jews, why would Frank use the word *ausrotten* ("to annihilate"), and why would he say that Germans should "rid themselves of pity"?

Shermer also read quotes from the diary of Joseph Goebbels, Hitler's minister of propaganda, which stated, "Jews should be liquidated; otherwise they will infect the population of civilized nations." Again Shermer argued that if the goal of the Nazis was deportation, why would Goebbels have used the word *liquidated*? Goebbels's diary also referred to a meeting with Adolf Hitler during which Hitler voiced a desire "to remorselessly eliminate the Jews. We must accelerate the process with cold brutality while doing an inestimable service to humanity." Goebbels's diary also stated that the Nazis wanted to put 40 percent of Jews in concentration camps to work and to kill the remaining 60 percent. Given that there were about eleven million Jews in Europe at the time, the estimate of six million deaths is remarkably close to what the Nazis had hoped to achieve. Shermer argued that it wasn't hard to figure out how many Jews had died during the Holocaust. Estimates, which range from 5.1 to 6.3 million, were derived by comparing the number of Jews reported to be living in Europe, the number transported to camps, the number liberated from camps, the number killed by the *Einsatzgruppen* (mobile police and SS units assigned to special missions in occupied territories), and the number alive after the war. Several historians have made these calculations, and all have come up with remarkably similar numbers.

Shermer then read from a speech given by *Reichsführer* Heinrich Himmler, one of the most powerful men in Nazi Germany. Himmler lamented that several German citizens had come to him asking that certain German Jews be spared. "We can talk about this [among ourselves] in an open manner," said Himmler, "but must never discuss this in public. I am now referring to the evacuation of the Jews and to the extermination of the Jewish people. This is something that is easily said. The Jewish people will be exterminated, sings every [Nazi] party member. This is very obvious. It is in our program—elimination of the Jews. But eighty million Germans come to us and each one has his decent Jew [whom they want to spare]. It is obvious that the others are pigs but this particular one is a splendid Jew." Again, argued Shermer, why did Himmler use the word *extermination* following *evacuation*?

Although Hitler never gave a direct order to kill Europe's Jews, following the Wannsee Conference, during which the details of the Final Solution were hashed out, Hitler remarked to his adjutants, "The Jew must clear out of Europe. Otherwise, no understanding will be possible among Europeans. It's the Jew who prevents everything. I restrict myself to telling them they must go away. But if they refuse to go voluntarily, I see no other solution but extermination." Again, the word *extermination*.

Shermer further argued that many of the Nazi war criminals testifying at the Nuremberg Trials told identical stories. All talked about how the gas chambers had been constructed and how Zyklon B (hydrogen cyanide) was introduced into the "showers" through chutes; they talked about the screaming, the piles of bodies with mouths hanging open, the difficulty of removing bodies from the chambers, and the particular revulsion some had about gassing women. Shermer also said that an inspection of the gas chambers, which had been located conveniently next to the crematoria, clearly revealed the chutes made for Zyklon B, the heating coils made to release the hydrogen cyanide gas, the "showers" with locks on the outside of doors rather than inside, and the ventilation system specifically designed to remove the gas after the killings had ended. Shermer talked about orders for large quantities of Zyklon B, photographs from Auschwitz of people being marched en masse to the gas chambers, photographs of Nazis burning bodies of the dead, and aerial photographs of images consistent with mass exterminations. What possible explanation could exist to explain all of this? Did the Holocaust deniers in the room believe that the Allies had constructed these chambers after they had liberated the camps?

Shermer said that while there was room for a revision of some of the smaller details about the concentration camps, there was no room for arguing whether the Holocaust happened. Shermer did, however, agree with Mark Weber on one point. The Nazis didn't initially plan to kill all of Europe's Jews; that plan evolved over time. When the Nazis found they could get away with depriving Jews of their citizenship, then prohibiting them from marrying non-Jewish Germans, then isolating them under intolerable conditions in ghettos, then killing them randomly on the streets, then deporting them against their will, they came to believe that they could get away with anything. That's why Hitler never issued a direct order to kill all the Jews at the beginning of the Third Reich. He didn't have to. "It doesn't matter what the Nazis originally intended," said Shermer. "It only matters what happened as the process went forward. They became bolder in their actions against the Jews. They did 'x' and found that they could get away with it. So now they could do 'x, y, and z.' Had it not gone that way, things might have been different." In support of the argument that the

Nazis' actions evolved into mass extermination, Shermer cited the work of Robert Jan van Pelt, an architectural historian who had written a book about his examination of Auschwitz. In his book, van Pelt describes how Auschwitz had been retrofitted for mass killings after the goals of the camp had changed.

Mark Weber tried to rebut Shermer's arguments, but couldn't. Instead he went on what can only be considered an anti-Semitic rant. He labeled Holocaust education a "campaign" by groups like the Anti-Defamation League and federal governments, decrying the self-importance of the United States Holocaust Memorial Museum in Washington, DC. He called the Jewish interest in preserving memory of the Holocaust arrogant and bigoted. Then he tried to make a moral equivalency argument Did Jews not also realize that twenty million Chinese died during World War II following attacks by the Japanese? Did they not realize that "gypsies and homosexuals," and other enemies of the German state had also died? Weber pointed his finger and raised his voice, clearly annoyed by Jewish people who regarded their lives as more important than those of other people. "We hear about the fate of one particular people during the war almost to the exclusion of everyone else," he said, referring dismissively to "this gas chamber business."

Shermer closed by saying that historic genocides were not unique to Germany, nor were they unique to Jews. But what made the Holocaust unique were the gas chambers, which were used in a systematic and calculated manner. Gas chambers next to crematoria next to mass graves—an assembly line of killing.

I learned a lot by watching Nye, Schwarcz, and Shermer debate the undebatable. For example, I didn't know about the six hundred eighty thousand cycles of winter and summer, the nine-thousand-year-old tree in Sweden, or the different methods of carbon dating the Earth. If I ever had to confront an evolution denialist, Bill Nye had armed me with those facts. Joe Schwarcz also taught me some things I didn't know—like that water changes its shape to conform to an active ingredient, but that this change doesn't last longer than a trillionth of a second;

and I didn't know the number of medical and scientific journals published daily. Now, when I am asked to deal with patients and families who want to use homeopathic remedies, I am better informed. Finally, Michael Shermer revealed quotes from Hans Frank, Joseph Goebbels, and Heinrich Himmler that I hadn't heard before. Also, I hadn't realized that the Holocaust had evolved and that Auschwitz wasn't initially constructed as a killing field.

What amazed me most about these three debates was how calm and reasoned Nye, Schwarcz, and Shermer had been— even under the most trying circumstances. Bill Nye had stood in front of hundreds of creationists at the Creation Museum. And Joe Schwarcz had been interrupted several times during his talk by audience members who were angry that he had challenged their beliefs about homeopathy. But the winner of the "Bravest Debater Award" goes to Michael Shermer, who stood in front of a group of old Holocaust deniers at an institute devoted to resurrecting the legacy of Adolf Hitler. Shermer must have felt like he was trapped in a scene from *The Boys from Brazil*. Also, each of these debaters had held their tempers for more than two hours. I, on the other hand, had expressed anger toward both the guest and the host during a television debate that had lasted only twenty minutes.

Nye, Schwarcz, and Shermer succeeded where I had failed because they weren't trying to convince the people they were debating (who weren't convincible) or the audience in front of them (who also weren't convincible); rather, they were trying to convince anyone watching their debate on television or YouTube. In the case of Bill Nye, that meant millions of viewers, and in the cases of Joe Schwarcz and Michael Shermer, tens of thousands. I, on the other hand, couldn't see beyond the two people in front of me, arguing with the guest that she had consistently misstated the facts, and with the host that she shouldn't have aired the debate to begin with. I didn't consider that some people watching the program might learn from some of the points I had made.

I offer one defense of my actions. I was myopic because I was angry. Every year several children are admitted to my hospital with severe infections that could have been prevented by vaccination. Invariably, this happens because parents had chosen not to vaccinate their children. And the reason they had made that choice was that they had read or heard bad information like that proffered by Mary Holland, who had been given a platform by Amy Goodman. Medicine is hard enough. There is much we don't know. But one thing we do know is that specific germs cause specific diseases and that at least some of these diseases can be safely and effectively prevented by vaccination. For example, recently a child was admitted to my hospital with bacterial meningitis caused by a strain of pneumococcus that could have been prevented by the pneumococcal vaccine. The parents, however, had refused vaccination in part because of what they had seen on television and read on the internet. The child's meningitis was so severe that her brain was pressing down on her brainstem, causing her to stop breathing. We intubated her and saved her life. But she will never see or walk or hear or speak again. That's the image that I have in my mind during these debates. And that's why I'm such a terrible debater.

So in the end, my advice is that debating the undebatable is worthwhile. But scientists (like me) need to take their focus off the host and the person they're debating and even the people in the room and shift it onto the people who aren't there. This is also true for debates on chat rooms and blogs on the internet. Don't think of these opportunities as debates. Think of them as teachable moments. I, however, will have trouble following this advice. It's just too emotional for me.

CHAPTER 6

Make 'Em Laugh

If you're going to tell people the truth, make them laugh.
Otherwise, they'll kill you.

—George Bernard Shaw

Growing up, I watched a lot of television. And no matter how farfetched the storyline, I always believed it. Some of my friends, however, were more skeptical, constantly looking behind the curtain, questioning, doubting.

Our conflicts started with *Adventures of Superman*, starring George Reeves. At issue was whether Superman could fly. My skeptical friends argued that this was impossible. Superman was heavier than air, and, although birds were also heavier than air, they had wings to lift themselves up. Patiently I explained that Superman was from another planet, Krypton. Things were different there. But they didn't buy it. Superman was a wingless, heavier-than-air man who should not have been able to fly simply by putting his arms out in front of him. They also didn't buy Superman's disguise as Clark Kent. Glasses? Really? Didn't Lois Lane, Perry White, and Jimmy Olsen notice that Clark Kent and Superman had the exact same face and build? *Adventures of Superman* aired from 1952 to 1958. My friends and I had these discussions when we were six years old.

When we were twelve years old, the argument shifted from *Adventures of Superman* to *The Patty Duke Show*, which ran from 1963 to 1966. The series featured two identical teenaged girls, Patty and Cathy Lane. Cathy was Patty's Scottish cousin. Despite their identical appearance, the girls were easily distinguished. They had different backgrounds: "Meet Cathy, who's been most everywhere from Zanzibar to Barclay Square, but Patty's only seen the sights a girl can see from Brooklyn Heights." And different personalities: "Where Cathy adores a minuet, the Ballet Russe, and crêpes Suzette, our Patty loves to rock and roll; a hot dog makes her lose control." They also dressed differently and had different accents and hairstyles. The sticking point for my friends was the notion of identical cousins. How could cousins be identical? They had different parents. The show explained this (to my satisfaction) by showing that the girls' fathers were also identical twins (played by the actor William Schallert). But what about the girls' mothers? Were they identical twins? And even if they were, wouldn't the chance of having identical twin cousins be more than a billion to one? I, however, remained convinced; the opening theme song was unequivocal: "Still, they're cousins, identical cousins, and you'll find they laugh alike, they walk alike; at times they even talk alike—you can lose your mind, when cousins are two of a kind." Good enough for me.

My first entrance into the world of skepticism came with the television show *Bewitched*, which ran from 1964 to 1972. *Bewitched* starred Elizabeth Montgomery as Samantha, a witch who marries a mortal man: Darrin Stephens, originally played by Dick York. Samantha vows to lead the life of a normal suburban housewife. Her family, however, disapproves of the mixed marriage—most notably, Endora, Samantha's mother, played by Agnes Moorehead. During the third year of the show, in 1966, the couple gave birth to a little girl, Tabitha. My problem wasn't that Samantha and Tabitha could make things appear and disappear by wiggling their noses. I believed that. My problem came in 1969 when Dick Sargent replaced Dick York in the role of Darrin. Suddenly, somebody else was walking through that door every night after work. And nobody said anything! Didn't Samantha notice that her husband was a different person, or Tabitha that she was sitting on another man's lap? Would somebody please offer an explanation? Maybe the writers could have had Samantha inadvertently turn Darrin into a newt. Or maybe Endora's disapproval of the marriage could have caused Darrin to finally pick up and leave. Something. Anything. I felt like *Bewitched* created a generation of children who worried that someday a stranger would replace their fathers and that no one would say anything. This marked the end of my "Age of Belief."

I don't know what happened to my skeptical friends. Maybe they became private detectives or Watergate investigators or FBI agents. Or maybe they became science bloggers: the single greatest force on the internet combating the dangerous misinformation proffered by science denialists. Blogs like David Gorski's *Respectful Insolence* and *Science-Based Medicine*, Todd W.'s *Harpocrates Speaks*, Matt Carey's *Left Brain/Right Brain*, and Michael Simpson's *Skeptical Raptor*, as well as podcasts like Steven Novella's *The Skeptics' Guide to the Universe* and David McRaney's *You Are Not So Smart*, among many others, constantly and reliably deconstruct claims made on anti-science websites.

Recently, another force has emerged to combat science denialists. Like the science bloggers, they, too, are skeptics; but unlike the bloggers, they do it for a living. Comedians. Al Franken, a *Saturday Night Live* veteran and later a United States senator, summed it up best: "Because I'd been a satirist since I was a teenager, I had plenty of experience identifying hypocrisies and absurdities." In other words, no one is better trained to appreciate the inconsistencies of anti-science rhetoric than comedians. And if you want to dismiss a notion as preposterous, nothing is more effective than making it a joke.

• • • •

COMEDIANS DO WHAT SCIENTISTS CAN'T DO. FOR EXAMPLE, WHEN I've been asked whether vaccines cause autism, I say, "Studies have shown that vaccinated children are not more likely to develop autism than unvaccinated children." Or, "Parents should be reassured that the evidence does not support the concern that vaccines cause autism." While accurate, these statements aren't particularly memorable. When Penn Jilette and his partner Teller addressed the same question on their Emmy Award–winning show *Penn & Teller: Bullshit!*, on the other hand, their answer was quite memorable. The episode aired on August 12, 2010.

Penn starts the show with a bang. "Hi, I'm Penn, and this is my partner, Teller. You may have heard that vaccination causes autism in one out of a hundred and ten children. Fuck that. Total bullshit. It doesn't." Penn says what many scientists and clinicians might think, but can never say—he's the raging id to our measured superego. (On the first episode of *Penn and Teller: Bullshit!*, Penn explained the cursing. "If one calls people liars and quacks, one can be sued," he said. "But *bullshit*, oddly, is safe. So forgive all the 'bullshit' language, but we're trying to talk about the truth without spending the rest of our lives in court.")

Penn then walks onto a stage featuring two groups of plastic bowling pins, about fifty on each side. Each pin represents a child. One group of bowling pins is protected by a large piece of Plexiglas; the other isn't. "We'll compare two groups of children," Penn says. (Teller rarely talks.) "Teller's group gets no vaccinations. My group does. I'll use this Plexiglas to represent the vaccinations. . . . In the 1920s, before the diphtheria vaccination was common, there were thirteen to fifteen thousand deaths a year from that disease. If you got it, your chances of dying were about 40 percent. In 1952, just before the Salk vaccine became common, there were about fifty-eight thousand cases of polio. If you got unlucky, you might end up permanently disabled or dead."

Penn now throws plastic balls at each side. The balls bounce off the Plexiglas, but knock down Teller's unprotected "children." Penn gets angrier, throwing the balls with greater force. "Meningitis, hepatitis A and B!" Penn yells. "Flu, mumps, whooping cough, pneumonia, rotavirus, rubella, smallpox, tetanus, chicken pox! Chicken pox! We have vaccinations against all of them! Which side do you want your child to stand on?" Penn, a magician, had magically presented hard statistics while at the same time keeping his viewers mesmerized.

Now Penn was really angry: "So even if vaccination did cause autism—which it fucking doesn't—anti-vaccination would still be bullshit!" This segment, lasting just one minute and thirty seconds, attracted more than 5.6 million views on YouTube.

Early in January 2011, I got a call from Emily Lazar, a producer for *The Colbert Report*. Would I be interested in coming on the show at the end of the month to talk about my new book, *Deadly Choices: How the Anti-Vaccine Movement Threatens Us All*? I had never seen the show, but my children assured me that this was one I shouldn't miss. They explained that a lot of young people get their news from *The Colbert Report* and *The Daily Show*, not from the major network news programs.

The show was scheduled to air on January 31. So I had the entire month to be nauseated and apprehensive. Although Lazar wouldn't tell me what questions I would be asked, she did direct me to certain episodes in which she felt the guests had done well or poorly—this way I could learn the do's and don'ts of *The Colbert Report*. Here's what I learned:

Lesson #1: Don't make jokes. *The Colbert Report* is a comedy show, and Stephen Colbert is the comedian. This means that you have to suppress your natural instinct to joke around with someone who is joking with you. Although you might think you're funny, you're not. People watched *The Colbert Report* to laugh at Stephen Colbert, often at his guests' expense. You're a straight man on a comedy show. Get used to it.

Lesson #2: Forget your talking points. One of the people Emily Lazar held up as a cautionary tale was Atul Gawande. Gawande is a best-selling author, a dramatic and engaging speaker, and a renowned surgeon at one of the nation's leading hospitals: Brigham and Women's Hospital in Boston, a Harvard Medical School teaching hospital. If Gawande was considered to have had a less-than-successful interview, then I was screwed.

The problem with his interview, which centered on his highly acclaimed book *The Checklist Manifesto: How to Get Things Right*, was that he kept coming back to his talking points. People didn't watch *The Colbert Report* to learn how hospitals could make fewer mistakes. They watched it to laugh at Stephen Colbert. I now understood that if anyone was to learn anything about vaccines during my interview, I should consider it a bonus.

Lesson #3: Stephen Colbert played a role. Colbert interviewed authors in front of a fireplace above which was the Latin phrase *Videri quam esse*. This is an inversion of the phrase, *Esse quam videri*, taken from Cicero's *On Friendship*. Cicero's phrase means "to be rather than to seem [to be]." Colbert had changed Cicero's phrase to mean "to seem to be rather than [to be]." He underlines this sentiment with his coined word *truthiness*, which he uses to imply that there are no truths. Anyone can claim anything. Stephen Colbert played the fool. For my interview, he played an aggressive anti-vaccine fool.

When I first arrived at *The Colbert Report* studio in Manhattan, I was escorted to a green room that contained great food that I was too nervous to eat. My wife and children, on the other hand, dug in. Although I hadn't expected it, Colbert met with each guest before the show. He explained to me that he played a character, that he was going to stay in character for the entire interview, and that, if the show was to be any good, I shouldn't let his character get away with anything. He said that he agreed with me completely. He said that his father—who I later learned had died in a plane crash when Colbert was ten years old—had been a doctor with an interest in infectious diseases and immunology. Colbert explained that an assistant would soon lead me to the set. Then he left.

Stephen Colbert did most of the show from a large desk shaped like the letter "C." When it was time to interview an author, he would jump up, wave to the crowd, and sit down opposite his guest at a small table. There was, however, about a five-minute period during which I sat at the table alone. It's hard to describe how frightening this is. In front of me was a live, cult-like studio audience chanting Colbert's name: "Stephen! Stephen! Stephen!" Because Colbert often makes fun of conservative Republicans, he receives a lot of hate mail. So much so that members of the studio audience must first walk through a metal detector before they sit down. During the show, two security guards constantly scanned the audience for unusual activity. The good news about having five minutes before Colbert walked over to the table was that this was plenty of time for my entire life to pass before me. While waiting, I read the first question written on a three-by-five-inch blue index card taped to Colbert's side of the table: "Why haven't you taken off your clothes for *Playboy*?" Then I looked over at my wife and children sitting expectantly in the front row.

Of interest, an odd sense of calm came over me when Colbert sat down for the interview. I felt like an astronaut about to blast off. No more preparation. No more practicing possible questions and answers. Everything was now out of my control. I was in the hands of Stephen Colbert, a man who had successfully humiliated people far smarter and cleverer than me—people like Alan Greenspan, a former chair of the Federal Reserve, and Amy Chua, the best-selling author of *Battle Hymn of the Tiger Mother*. My wife and I have a name for this how-the-hell-did-my-career-end-up-here moment. We call it the Iowa-cow-pasture moment.

In the late 1990s, I collaborated with a research group interested in veterinary vaccines. Our lab had been working on a method to protect viral vaccines from the harsh environment of the stomach. The goal was to give more vaccines by mouth, to avoid all those shots. The animal-testing facility was located in the middle of a cow pasture in Livingston, Iowa. To keep the animals safe, the facility had a shower-in, shower-out rule. The showers, however, were *outside* the facility. So, there I was, wondering how four years of medical school, three years of pediatric residency, three years of a pediatric infectious disease fellowship, and many years of basic science research funded by the National Institutes of Health had led me to standing naked in the middle of a cow pasture in Iowa.

Then my segment started. "Alright, I'll bite," Colbert starts off. "Why should I vaccinate my kids?"

Great. A home run pitch. No mention of taking off my clothes for *Playboy*. Maybe this wasn't going to be as bad as I thought. "Because vaccines have saved their lives," I say.

Then Colbert systematically parodies the reasons that some parents have chosen not to vaccinate their children; first, vaccines are victims of their own success. "I had my kids immunized against rubella," says Colbert. "Guess what? They never got rubella. It was a waste of money."

"See, that's the thing about vaccines," I said. "When they work, absolutely nothing happens. But that's a good thing, right?"

"I don't know, you see," says Colbert, "because I am entirely motivated by fear, and if I don't see something around me to fear, I think a little outbreak [is] not a bad idea. . . . But if I was afraid, I would do it."

"Well, people are getting sick, and they are dying," I say.

"Says you!" Colbert counters.

This was a little intimidating. But as Colbert had instructed me before the show, I tried to separate the person Stephen Colbert from the character he was playing and fight back. "We have a whooping cough outbreak in California that's bigger than anything we've seen in more than fifty years, and there are ten children who have died of whooping cough. So a choice not to get a vaccine is a risky choice."

Colbert next addresses the issue of social responsibility. "But if everyone around me is immunized," he says, "then why do I or my kids need to get immunized? We're protected by them, right?"

"The problem is that there are about five hundred thousand people in the United States who can't be vaccinated. . . . They depend on those around them to protect them." Long, long pause. "And that's you," I say, pointing to Colbert. (Some of my friends later commented on how well they thought I had done at that moment—that I had somehow stumped Stephen Colbert.

But Colbert was playing a role. He only pretended to be stumped.)

"OK," he continues, "so can't we put those people in, like, the boy in the plastic bubble? Can't we just do that?"

"So, like, a sort of big bubble for five hundred thousand people?" I ask.

"One at a time," Colbert says. "Let's be practical here."

Then Colbert addresses the most outrageous claim made by anti-vaccine parents: that natural infections are a good thing. "Shouldn't we just put sick people together and build up natural immunities the way the cavemen did?" Colbert asks.

"And that's what we used to do," I say. "And what would happen is that every year you would have thousands of children dying from measles or whooping cough or having congenital birth defects from rubella or being paralyzed by polio. Fortunately, we don't have to do that anymore."

"See, now, this isn't fair," he says, "because you're playing the children-dying card. How am I supposed to fight that? Let's keep this intellectual."

Colbert closes the segment with a joke. "Were you immunized as a child?" he asks.

"Yes," I say.

"Do you have the little scratch on your arm [that] looks like an asterisk?"

"Yes, that's the one I got," I answer.

"Now I understand, smallpox, obviously bad," he says. "But swimsuit season is coming up, and you've got a scar on your arm. Isn't there a better way to do that?"

"That scar seems to be a fairly small price to pay during swimsuit season," I say.

"Depends on where it is," he says.

Then he reached out to shake my hand, and it was over. The segment lasted about five minutes, but it seemed like thirty seconds. Colbert took the time to speak with my kids before we left. They were in heaven.

I never would have imagined that Colbert would invite me on to his show one more time and that the next time I would be booed.

On April 29, 2014, I appeared on *The Colbert Report* again. "Here to tell us we're all going to die is the director of the Vaccine Education Center at the Children's Hospital of Philadelphia," said Colbert. And we were off.

On this segment, Colbert took on the issue of cause and effect, brilliantly: "You have to admit that the amount of vaccines given to young children each year increased in the '90s, and the diagnosis of autism rose at the same time," he says. "That's a corollary effect, OK? It's the same way that the iPhone is introduced and World War II vets start dying in the same decade. They've got to have something to do with each other."

"You know, it's perfectly reasonable for the parent to ask the question," I say. "My child was fine, they got a vaccine, and now they aren't fine. Could the vaccine have done it? It's an answerable question, and it has been answered again and again and again. The question is, Why do 29 percent of Americans still think vaccines [cause autism] when they've been shown not to?"

Colbert ends the segment with a direct challenge: "What if I were to tell you you sound like you're in the pocket of Big Pharma? Big Pharma makes these inoculations. They're making money. They're greasing your palm. That's out there. Now, respond to it."

"I'm not in the pocket of Big Pharma," I say. "I'm in the pocket of children. That's why I work so hard to educate people about vaccines. I want to protect children."

The entire studio audience boos, loudly. I don't know what I had said that was so wrong. But everyone else does, including Colbert. He's great, though. He leans across the table and, stepping out of character, says, "Let's do that one again. And we'll go out with this question."

So I answer the question one more time: "You can't on the one hand claim that vaccines have saved our lives and have allowed our children to live longer, better, healthier lives, and then just dismiss the people that make vaccines so safely and so effectively." Then I do something that you are specifically told not to do. I look directly at the studio audience and ask, "Was that any better?"

They cheer. Again, Colbert had let me win. "Alright," he says, "if you want to play the longer, better, healthier lives card, there's nothing I can do."

The Colbert Report was live to tape, meaning that the final version would be edited. People who watched the show on television never saw my first answer or heard the booing. They only saw my second answer followed by loud cheers. (My question to the audience was also edited out.) So basically you see someone saying something nice about vaccine makers followed by loud cheering. No doubt a television first.

While I was walking backstage after the interview, I asked the assistant producer why people had booed. She said that guests often forget that this is a comedy show, that when I had said that I was in the pocket of children, I had sounded like a pedophile. I can honestly say that that interpretation had never crossed my mind. But I haven't used that phrase again. On the way back to the train station, I asked my wife and daughter whether they would ever have imagined that my first answer could have made me sound like a pedophile. "Yes," said our daughter. "I thought of that immediately. That's why I booed." ("You booed, honey?")

Two months later, on June 4, 2014, a producer for Jon Stewart asked me if I would be willing to go on *The Daily Show*. The correspondent would be Samantha Bee. Where Penn Jillette had screamed at anti-vaccine activists and Stephen Colbert had pretended to be one, Samantha Bee ridiculed one to her face. Further, Bee treated her guest as if her beliefs weren't only wrong, but dangerously contagious—as if the guest, herself, was a virus.

Before the interview started, Samantha Bee met me in the research building where I work. I actually didn't know who she was (which is consistent with my children's belief that I'm a cultural idiot). Bee was humble, which was surprising given that she has been a major force in the "boys' club" of late-night talk shows. She introduced herself assuming that I didn't know who she was. (When I first met her, I thought she was the producer.) She seemed painfully quiet and shy. (I can read women. It's hard to explain. It's like a gift.) Samantha Bee—as everyone on the planet except me, apparently, knows—is a wild woman, as evidenced by what she was about to do during our interview and by the title of her current show, *Full Frontal with Samantha Bee.*

Bee confronted science denialism head on, starting her segment showing a series of interviews from Fox News during which several conservative pundits offered their views.

Addressing the audience, Bee says, "This right-wing science denial has tragic, real-world consequences."

"When you cross the line into being a scientific denialist," I say, "then you have the potential to do a lot of harm."

"Is it really fair to blame the other side, though?" asks Bee. "I mean, every scientific fact has a counter-fact that is true for other people."

"The good news about vaccines," I say, "[is that] they're not a belief system. They're an evidence-based system."

Bee then shows a graph of the incidence of various diseases before and after the introduction of vaccines. "He's right," she says. "Because of these right wing nut jobs, outbreaks of vaccine-preventable diseases are occurring in the red states of California, New York, and Oregon. Wait, what the fuck is going on here?"

I'm asked to explain why vaccine denialism appears to be a blue-state problem, given that the conservative science denialists on Fox News presumably represent red states. "There are communities that have large populations of Caucasian, upper-middle-class residents who are college educated, often graduate-school educated, who believe that by simply Googling the word *vaccines* on the internet, they can know just as much [as], if not more than, anyone who's giving them advice."

"It's happening in my community?" says Bee. "People who juice?"

Bee then sets out to stop the spread of bad information about vaccines. "I had to find the source. How about this lifestyle blogger, Sarah Pope?"

"These vaccines are loaded with toxins," says Pope. "And I'm not going to inject them into my kids. Period."

Bee next takes on the issue of shared responsibility, letting Pope hang herself with her own logic. "There is no herd immunity from vaccination," says Pope. "That's a myth . . . I believe that the decline in epidemics is due to other reasons besides vaccination. Getting the filthiness of the horses out of the streets."

"You know that you don't get measles from horseshit, right?" asks Bee, worried about how many people are actually listening to Pope.

Pope then says that her blog currently has about forty-six thousand subscribers.

The producers of *The Daily Show* bring two cameras to these interviews. The next scene explains why.

After Pope states her blog's number of subscribers, Bee literally foams at the mouth. Copious amounts of thick, white foam pour out. Pope is disgusted.

"I'm so sorry," says Bee, holding up a large white tablet. "I thought this Alka-Seltzer was just a giant breath mint."

(One purpose of the second camera is to catch guests' expressions when Samantha Bee does something unexpected. During my interview, without warning, Bee threw her hands to her head, jumped up, screamed "What the fuck?" and ran out of the room. I suspect that my face registered something between shock and horror.)

Bee then sets out to prevent the misinformation spread by Sarah Pope from going viral—treating Pope as if she herself is a virus. "I had to trace the spread of this left-leaning idiocy," says Bee. "It starts on blogs like Pope's, goes viral on Twitter, and then replicates wherever progressives congregate." (Bee shows a picture of a Whole Foods store.) "And then, when it jumps hosts into a celebrity, it goes airborne. You can catch it from an iPhone." (Bee shows a photo of herself putting an iPhone into a hazmat bag.) "Or over soy lattes. I had to stop it."

"Do you remember where you last had tea?" Bee asks Pope, apparently panicked that it might be too late to stop the viral spread.

"It was Starbucks," says Pope.

"Which Starbucks?" screams Bee.

Back to me in my office: "I think, sadly, that the only way this gets better is when we see more and more outbreaks," I say.

"So the cure for the anti-vaccination epidemic," says Bee, "is observing the very real negative effects of the anti-vaccination movement?"

The segment ends with pictures of ice caps melting and children in iron lungs. "So there *is* a cure for science denial," says Bee. "Once Florida is underwater and we all have polio, it'll be better."

At one point during the filming of my interview for *The Daily Show*, Samantha Bee asked if she could lie down on top of a workbench in my laboratory. She wanted me to read to her from the textbook *Vaccines*, which is about sixteen hundred pages long, as if I were telling her a bedtime story. At the same time, she wanted me to gently pat her head. So I did it. It was another Iowa-cow-pasture moment, but it was worth it. Nothing dismisses ill-founded beliefs better than humor. And although some of this interview had been uncomfortable for me, I knew that I had to be willing to set aside the small amount of residual pride that might have been left after surviving twenty-five years of the NIH granting process and just do it.

I wondered how Sarah Pope would handle being publicly ridiculed on national television. She loved it. On her blog the next day, Pope wrote about how she had made her points and how anti-vaccine activists had praised her for her bravery. She later called Samantha Bee to thank her. To Sarah Pope, there was no losing.

On May 11, 2014, on his show *Last Week Tonight*, John Oliver took on climate change denialism.

Like evolution, homeopathy, and vaccine safety, climate change is also not a matter of debate. Owing to the unchecked burning of fossil fuels, the level of carbon dioxide in the Earth's atmosphere has risen dramatically. Because carbon dioxide is a heat-trapping gas, the Earth's surface temperature has also been increasing. The most obvious outcome has been a series of massive storms and devastating heat waves, like those affecting Europe, India, and Pakistan. In the United States, the number of days with record-high temperatures has doubled during the past fifty years.

Increased temperatures have caused a melting of Arctic ice sheets, resulting in a ten-inch rise in ocean levels: a phenomenon that has already caused problems for low-lying nations in the tropical Pacific. If this warming trend continues, unusual diseases will begin to appear in unusual places; mosquito-borne illnesses like dengue, malaria, chikungunya, and Zika will spread beyond tropical regions. Zika virus has already traveled up into Florida and Texas, resulting in more than two hundred cases in 2016.

Today, about thirty billion tons of carbon dioxide are released into the atmosphere every year. If we don't do something now, by midcentury it will be too late. Typically, scientists divide the Earth's geological history into a series of eras, like the Mesozoic and Paleozoic. The current era is the Anthropocene, meaning that the largest threat to the planet isn't meteors or tectonic shifts in land masses; it's human activity. Despite clear evidence and virtual universal agreement among scientists, a University of Texas poll found that 24 percent of Americans weren't convinced that climate change was real.

"The Earth," says John Oliver. "You may know it as that blue thing Bruce Willis is always trying to save. Or from its famous collaboration with Wind and Fire. Or just simply as that place where George Clooney lives. Anyway, the Earth had some genuinely bad news this week."

A woman in voice-over then states that the White House recently reported that global warming threatens every part of the United States.

"This isn't something in the distant future," says President Obama in a clip. "Climate change is already affecting us now."

In another clip, an MSNBC commentator says, "There's that Gallup poll that came out last month, which found one in four Americans is skeptical of all the effects of climate change and thinks this issue has been exaggerated."

Scientists will often say that the only thing that matters is what the data show. John Oliver makes this same point, but far more effectively. In response to the report that 25 percent of Americans don't believe in climate change, he says

Who gives a shit? That doesn't matter. You don't need people's opinions on a fact! You might as well have a poll asking which number is bigger, fifteen or five? Or, do owls exist? Or, are there hats? The debate on climate change should not be whether or not it exists; it's what we should do about it. There is a mountain of research on this topic. Global temperatures are rising. Heat waves are becoming more common. Sea surface temperatures are also rising. Glaciers are melting. And of course, no climate change report is complete without the obligatory photo of a polar bear balancing on a piece of ice."

(Oliver then shows the obligatory photo.) Oliver continues,

The only accurate way to report that one out of four Americans are skeptical of global warming is to say a poll finds that one out of four Americans are wrong about something. Because a survey of thousands of scientific papers that took a position on climate change found that 97 percent endorsed the position that humans are causing global warming. And I think I know why people think this issue is still open to debate: because on TV, it is. And it's always one person for, one person against, and it's usually the same person for.

Oliver then shows clips of Bill Nye appearing on several television talk shows. "More often than not, it's Bill Nye the Science Guy versus some dude. And when you look at the screen, it's fifty–fifty, which is inherently misleading. If there has to be a debate about the reality of climate change—and there doesn't—then there is only one mathematically fair way to do it."

Oliver then brings onto the stage ninety-seven people dressed as scientists, including the real scientist Bill Nye, to represent the 97 percent of scientists who support the data on climate change, and three more people to represent the 3 percent of scientists who don't believe in climate change. Oliver encourages the two sides to engage in a debate, which quickly devolves into a shouting match in which no one can be heard. Oliver concludes, "This whole debate should not have happened." Oliver's takedown of false balance went viral—more than seven million people have viewed this segment on YouTube.

Climate change wasn't the only issue that John Oliver addressed on *Last Week Tonight*. On June 22, 2014, he took on Mehmet Oz and his promotion of "miracle" dietary supplements. He warned viewers about what can happen when health care products are unregulated.

"Dr. Oz has become one of America's most trusted docs," says a commentator in a clip. "But on Capitol Hill on Tuesday, he was on the hot seat."

The clip then shows a Senate hearing, at which Oz was being questioned regarding the marketing of dietary supplements, many of which he had repeatedly endorsed on *The Dr. Oz Show*.

"Oz's advice is so influential that one mention of a product can cause sales to skyrocket," says a reporter in another clip, referring to the phenomenon as the "Dr. Oz effect." Citing an example, the reporter states, "After Dr. Oz touted a substance called green coffee bean extract, one company in Florida sold half a million bottles."

Enter John Oliver. "What's so wrong with that?" he says. "Name me one case where a man named Oz claimed mystical powers and led people horribly astray! Name me one case!

You can't do it! The only problem with the Dr. Oz effect is that magic pills don't technically exist, and Dr. Oz knows that."

Oliver then shows a clip of Oz hesitating in the Senate hearing when asked whether he believes in magic weight-loss products. Next up is a clip in which Oz claims that green coffee bean extract could cause dramatic weight loss. "See?" says Oliver. "He never said there was a magic pill. He said there was a magic bean. That's clearly entirely different. Because magic beans are a very real thing that you trade your cow for so you can steal a golden harp from a giant. That's science!"

In front of the Senate panel, Oz says, "I recognize that oftentimes [the products I promote] don't have the scientific muster to present as fact."

Oliver is astonished. "But that's the whole point," he says. "You're presenting it *as a doctor.* If you want to keep spouting this bullshit, that's fine. But don't call your show *Dr. Oz*; call it *Check This Shit Out with Some Guy Named Mehmet.*"

Oliver continues his rant. After discussing dietary supplement disasters like OxyElite Pro (which caused more than fifty people to suffer hepatitis and acute liver failure) and L-tryptophan (which killed thirty-eight people and caused thousands to suffer neurological symptoms), Oliver says, "Dr. Oz is just a symptom of the problem. The disease is the fact that dietary supplements in the U.S. are shockingly unregulated. . . . The FDA has little authority to investigate the contents of supplements until people are already getting sick from them. . . . The industry is essentially supposed to police itself. That's like one of those porn sites that asks you to enter your own age. Which basically just ends up teaching children how to subtract eighteen from the current year."

Oliver then says that Oz's problem is that he has done too many shows, forcing him to go beyond the bounds of scientific credulity. "You're going to get tempted to overstretch," says Oliver, "and do a show with a title like 'Can an Aspirin a Day Keep Cancer Away?'—to which the answer is clearly no. Because I feel like if it did, we wouldn't be hearing about it for the first time at 4:00 pm on a Wednesday afternoon. 'Hey, did you hear that they cured cancer?' 'No, I was watching Wendy Williams.'"

Because the dietary supplement industry has annual revenues of about $32 billion, few in the mainstream media had been willing to take it on. But John Oliver did. This segment also went viral, garnering more than twelve million views on YouTube.

On February 27, 2015, Jimmy Kimmel became the next comedian to take on the anti-science movement. The timing wasn't coincidental. Neither was the location of Kimmel's show. Penn and Teller filmed in Las Vegas. Both *The Colbert Report* and *The Daily Show* were filmed in New York City. But *Jimmy Kimmel Live!* was filmed in Los Angeles. In early 2015, Southern California was at the epicenter of what was to become a massive measles epidemic caused mostly by parents who had chosen not to vaccinate their children.

First, Kimmel addresses the issues of relative risk and dismissal of expertise. "Here's another area which I feel like we're headed in the wrong direction," Kimmel begins:

I've been hearing a lot of talk lately. And I don't know if this is more prevalent in L.A. than in other places. I feel like it probably is. But there is a small but still sizable group of people who are choosing not to vaccinate their children. Here in L.A., there are schools in which 20 percent of the students aren't vaccinated because parents here are more scared of gluten than they are of smallpox. And, as a result, we now have measles again. . . . I know if you're one of these anti-vaccine people, you probably aren't going to take medical advice from a talk show host. And I don't expect you to. I wouldn't, either. But I would expect you to take medical advice from almost *every doctor in the world*. See, the thing about doctors is they didn't learn about the human body from their friend's Facebook page. They went to medical school where they studied all sorts of amazing things like how to magically prevent children from contracting horrible diseases by giving them little shots. You know those little shots of Botox—which is botulism by the way—that you get in your face to make your head look smooth and your eyes look crazy? A little shot like that and, poof, polio is gone. But some people do not buy into that because they did a Google search, and Jenny McCarthy popped up. And she had clothes on, so they listened to what she had to say and decided not to vaccinate their kids. . . . Oh, by the way, you should also let your kids smoke. Why wouldn't you? The only people who say not to are doctors, and they don't know. So, I feel like this is starting to snowball. So we invited some real doctors to address this. Again, these are not actors. They are actual medical professionals. Every one of them is a real doctor. So hear them out, and then decide for yourself.

Kimmel then presents a clip of a number of doctors expressing their frustration with parents who choose to leave their children vulnerable:

"I thought we settled this in the '50s."

"Hey, remember that time you got polio? No, you don't, because your parents got you fucking vaccinated."

"I did four years of undergrad, four years of medical school."

"I had to go to school for eight god-damned years."

"Now I have to use my only day off to talk to you idiots about vaccines."

Kimmel had taken Penn Jillette's anger to another level. Now, it wasn't a comedian who was cursing at parents. It was doctors. And believe me, they weren't acting. Medicine is frustrating enough without having to stand back and watch parents make bad choices based on bad information that has caused far too many children to suffer. And it's all perfectly legal. If parents don't want to vaccinate their children, all they have to do is claim that they have a religious reason (forty-seven states have religious exemptions to vaccinations) or a philosophical reason (seventeen states have philosophical exemptions). And vaccines are just the tip of the iceberg. Parents can legally choose prayer instead of antibiotics for meningitis, or instead of insulin for diabetes, or instead of chemotherapy for treatable cancers. Most states have religious exemptions to child abuse and neglect laws, including medical neglect. You just want to scream. But the simple truth is that running through the heart of American jurisprudence is a libertarian streak that is wide and deep, especially in the family realm. We simply don't like to tell parents how to raise their children, even when those children are put in harm's way unnecessarily.

Less than a week after his first go at the anti-vaccine crowd, Jimmy Kimmel did something that made him either the bravest or most reckless man confronting science denialism. He took the notion of parental choice about vaccination to its illogical end. He recruited two staffers, Jack and Becky, to march down the streets of Los Angeles holding signs and chanting, "A Child's Right to Choose! A Child's Right to Choose!" (Jack and Becky had previously tried to get people to sign a petition to ban the word *peanuts* from the song "Take Me Out to the Ball Game," supposedly because their child was allergic to peanuts.)

The two interact with a number of passersby and show them a video titled "Pediatric Vaccine Study." The video shows a man, the supposed researcher of the vaccine study, holding a syringe with an unsheathed needle in front of children five to seven years old, the "research subjects," and asking them if they want to get a vaccine.

A somber voice-over gives the results: "We found that 100 percent of respondents said they would prefer not to receive a vaccination shot."

In the next experiment, the researcher gives the children a choice. They can either get a vaccine or a lollipop. The result: "We found that 100 percent of respondents preferred the lollipop to the vaccination shot."

Kimmel was arguing that if you're going to allow parents to make an uninformed choice, you might as well let children do it, too. One can only imagine the hate mail he received after this segment.

CHAPTER 7

Science Goes to the Movies

Victor Moritz: "You're crazy!"
Dr. Henry Frankenstein: "Crazy, am I?
We'll see whether I'm crazy or not."

—FRANKENSTEIN (1931)

The entertainment industry educates the public about science in a different venue: movies, which not only shape opinion, but reflect some of our worst fears about science and scientists— a fear best crystallized in a single icon: Frankenstein. The 1931 movie, based on the book of the same name written by Mary Shelley in 1818, tells the story of a scientist, Dr. Henry Frankenstein, who robs graves to procure various organs and limbs from the dead. Unfortunately, his creation, brought to life with an electrical current, was made using the brain of a murderer. Frankenstein, a literal monster created by a scientist, terrorizes the countryside.

Modern-day Frankensteins are created through something scientists call gain-of-function studies. Indeed, many scientists have now banded together to oppose them. In gain-of-function studies, microbes are altered to enhance their contagiousness. While these studies can be instructive, they also have the power to unleash uncontrollable diseases. For example, measles, mumps, chicken pox, and influenza viruses are highly contagious. Rabies, on the other hand, isn't. Unlike these other viruses, rabies is acquired from the bite of an infected animal, not from unseen droplets created by coughing. If the rabies virus was reengineered such that, like the respiratory viruses, it could be spread more easily, the result would be a highly contagious, uniformly fatal infection.

The gain-of-function phenomenon has been the subject of two movies: *Outbreak* (1995) and *Contagion* (2011). One was remarkable in its ability to educate the public about science; the other wasn't. The difference between the two shows how we can be influenced to develop reasonable or unreasonable expectations about how infections are spread, how vaccines are developed, and how outbreaks are controlled.

We'll start with the movie that got the science right.

• • • •

CONTAGION OPENS WITH A BLACK SCREEN AND THE SOUND OF A woman coughing. When the black screen dissolves, Gwyneth Paltrow is sitting in an airport bar in Chicago having just returned from a trip to Hong Kong. On-screen text reads "Day 2." Paltrow is talking on a cellphone to a man with whom she has recently had sex. She is on her way back to Minneapolis, where she lives with her husband and son. The encounter with the man in Chicago, apparently, was an extramarital affair. After Paltrow coughs into her hand, the camera lingers on everything she touches: a bowl of peanuts, her credit card, the cash register.

Back in Hong Kong, another man is coughing in a subway car. On-screen text reads "Kowloon Hong Kong, population 2.1 million." The camera lingers on a railing he has just touched. Dazed and disoriented, the man wanders onto a busy street, where he is struck and killed by a truck—but not before he gets onto an elevator full of people, the camera resting on the elevator buttons. Another man traveling by plane from Hong Kong arrives in "Tokyo, population 36.6 million," where he collapses, seizes, froths at the mouth, and dies.

The scene shifts to "London, population 8.6 million," as a sick woman walks into an office of graphic designers; the camera lingers on a notebook she places on her desk. In the next scene, she is lying face down in a hotel room, dead. The people who died in Hong Kong, London, and Tokyo will soon be linked to a casino in Hong Kong—each having been exposed to Paltrow.

Paltrow returns to "Minneapolis, population 3.3 million," where she hugs her husband (played by Matt Damon) and their son. The next day, Damon picks up their son from school, where he has been coughing. The camera lingers on a door handle.

"Day 4." Paltrow collapses in her kitchen, seizing uncontrollably. Damon rushes her to the emergency room, where, despite all efforts, she dies. Damon then receives a frantic call from his son's babysitter; he comes home to find that his son, too, has died. Because Paltrow hugged her son on "Day 2," the incubation period of the disease—the time between first exposure and the onset of symptoms—is two days. We also learn that once symptoms develop, death is rapid and inevitable.

On "Day 5," we see the man in "Chicago, population 9.2 million," with whom Paltrow has had an affair, being wheeled out of his home by paramedics, his wife anxiously running by his side. He, too, will soon die. The scene shifts to a morgue, where a pathologist pulls Paltrow's scalp down in front of her face and removes the top of her skull. After exposing the brain and finding severe inflammation (encephalitis), the pathologist steps back in horror, certain that he has just been exposed to a deadly virus. The pathologist calls the CDC.

Kate Winslet plays the CDC epidemiologist in charge of the outbreak. Winslet predicts that the virus is spread by tiny respiratory droplets, as well as by fomites: objects that an infected person might touch, like handrails, elevator buttons, and ATMs. The writers of *Contagion* then do something that no other movie about epidemics has ever done: They explain the concept of contagiousness.

On a blackboard in front of members of the Minnesota Department of Public Health, Winslet writes "R_0," explaining that the "R" stands for "reproducibility index." She writes "Smallpox: 3; Influenza: 1; Polio 4–6." Then she explains that someone infected with smallpox will infect three more people every day. Winslet estimates that—given what is known about the outbreaks in Tokyo, Hong Kong, London, and Chicago—the R_0 for this virus is two. The audience now understands how two infected people can quickly become four, eight, sixteen, thirty-two, sixty-four, and so on. Given that the incubation period is short, that the disease is invariably fatal, that no antiviral drugs are available to treat it, and that no vaccine exists to prevent it, the viewer also understands that this epidemic is unstoppable.

The scene shifts back to the CDC, where *Contagion* again distinguishes itself as the only blockbuster movie to describe the science of viruses and viral vaccines carefully and accurately. A CDC virologist, played by Jennifer Ehle, has now isolated and defined the genetic sequence of the epidemic virus, which she calls MEV-1. Ehle explains that MEV-1 is a combination of genetic sequences from two viruses: one a pig virus, the other a bat virus. Bat viruses, which can cause fatal encephalitis, aren't spread from one person to another; they are spread only from bats to people. However, by incorporating genetic sequences from a pig virus, this new combination virus can now be spread easily by coughing, sneezing, or even talking. (Ehle's explanation is entirely plausible. Pigs and humans share the same viral receptors on cells that line the nose and throat. Again, the writers of *Contagion* advance a sophisticated concept, accurately.) "Somewhere in the world, the wrong pig met up with the wrong bat," says Ehle.

The next day, the CDC estimates that the MEV-1 virus has killed thousands of people and is on its way to killing hundreds of thousands. The writers of *Contagion* again break new ground by taking time away from usual movie fare (sex and car-chase scenes) to explain how vaccines are made. Ehle says that you could kill the virus (the way that Jonas Salk made his polio vaccine), weaken the virus (the way that the measles, mumps, rubella and chicken pox vaccines are made), or take just one of the deadly virus's genes and clone it into a different, harmless virus (the way that the Ebola and dengue vaccines are made). The MEV-1 virus, however, is so deadly that it immediately kills every type of cell in which it has been grown, making it impossible to weaken in the laboratory. This is another sophisticated concept. But the writers trusted the audience to understand it. Then Kate Winslet, the CDC investigator, dies from the disease (which seemed horribly unfair since she had only recently survived the sinking of the RMS *Titanic*).

The breakthrough comes when a virologist in San Francisco, played by Elliott Gould, finds that he can grow MEV-1 in bat lung cells that he obtained from a collaborator in Geelong, Australia. This allows Ehle to weaken the virus.

To test her vaccine, Ehle inoculates experimental monkeys and then challenges them with MEV-1. If the vaccine works, the monkeys will live; if not, they will die. The next several scenes show researchers stuffing dead monkeys into plastic bags. On the fifty-seventh attempt, however, the monkeys survive. Again, give credit to the writers for showing so many failed attempts. The scientific advisor for *Contagion* was Ian Lipkin, a professor of epidemiology at Columbia University and an expert in the field of unusual viruses coming from unusual places. In tribute to Dr. Lipkin, the Elliott Gould character is also named Ian.

In the midst of the pandemic, society breaks down. Schools and churches empty. Banks are overrun. Grocery stores are looted. Police and fire departments disband. Trash accumulates. Airports close. The president of the United States is taken to an undisclosed location, and Congress works underground. The death toll reaches 2.5 million.

The writers of *Contagion* take on one more challenge that makes this film not only accurate but brave. They include a character, played by Jude Law, who touts a homeopathic cure for MEV-1 called Forsythia. On his blog (*TruthSerumNow*), which has more than two million followers, Law shows himself coughing and haggard, pretending that he, too, is now infected. Then he takes Forsythia. During the next few days, Law appears to recover. Convinced that a cure now exists, people break into pharmacies trying to get Forsythia. Law, who makes $4.5 million from his bogus cure, is a direct slap in the face of dietary supplement hucksters.

The action shifts back to the CDC, where Ehle explains how the MEV-1 vaccine will be tested, mass produced, and distributed, again accurately. In response, Law appears on CNN decrying scientists who have urged vaccination. Law argues that the vaccine hasn't been tested long enough and that we shouldn't trust government and pharmaceutical company scientists—a common refrain from anti-vaccine activists. When Law is later found to have faked his illness, he's arrested for securities fraud, conspiracy, and manslaughter. While this might seem over the top, consider that homeopathic doctors in Canada sell nosodes: homeopathic vaccines. Homeopaths make these vaccines by taking a virus and diluting it in water to the point that it's not there anymore. In other words, they're selling water as a vaccine. Given that some people could suffer and die from vaccine-preventable diseases as a consequence, these homeopaths could reasonably be charged with the same crimes as the Jude Law character in *Contagion*.

Unlike *Contagion*, *Outbreak*, also featuring the potential dangers of gain-of-function studies, sacrificed accuracy for drama.

Outbreak opens in a war zone in Zaire. The year is 1967. Men in hazmat suits are walking through a village where an unknown illness has killed forty-eight people. Many others are dying, blood coming out of their mouths and eyes. The men in hazmat suits assure those who are dying that everything possible will be done to save them. A few hours later, a plane is seen flying overhead; the men rejoice, assuming that medicines and supplies are on their way. Joy turns to horror when they realize that the parachute isn't carrying supplies; it's carrying a bomb that destroys the entire village. Monkeys run from the wreckage.

Thirty years pass.

Next we learn that the United States is preparing to use this deadly African virus as part of a germ-warfare program. (Here, art imitates life. Both the United States and Russia have developed programs to put plague bacteria and hemor-rhagic fever viruses into bombs and missiles. Richard Nixon ended the biological weapons program in 1969 and ordered all existing stockpiles destroyed.)

The scene shifts to the U.S. Army Medical Research Institute of Infectious Diseases in Fort Detrick, Maryland, where bio-safety hazards are explained:

- Level 1, mild risk: pneumococcus, salmonella.
- Level 2, moderate risk: hepatitis, influenza, Lyme disease.
- Level 3, high risk: anthrax, HIV, typhus; vaccinations required (which would have made a lot more sense if vaccines actually existed to prevent HIV and typhus).
- Level 4, extreme biohazard, maximum security: Ebola virus, Hantavirus, Lassa virus; no known cures or vaccines.

Soon we learn that another outbreak has occurred in Zaire. Several scientists, played by Morgan Freeman, Cuba Gooding Jr., Dustin Hoffman, Rene Russo (Hoffman's ex-wife in the movie), and Kevin Spacey, are dispatched to control the out-break and find its cause. "Looks like we have a Level 4," says Freeman. "I'm flying to Zaire."

The scientists arrive to find that the symptoms are the same as those in the 1967 outbreak: bleeding from the eyes, ears, mouth, and rectum, accompanied by liquefaction of internal organs—in short, a hemorrhagic fever virus like Ebola. (For the record, hemorrhagic fever viruses don't actually liquefy internal organs. *Outbreak* is based loosely on Richard Preston's best-selling novel *The Hot Zone*, which tells the story of the Ebola virus, a deadly pathogen that first appeared in Zaire in 1976, killing three hundred people.) Bodies are lined up in rows, wait-ing to be burned in one of several fires dotting the village. One child is riddled with blisters, crying, and sick—his two dead parents lying beside him. Seeing this, Cuba Gooding Jr. vomits into his hazmat suit. Because the outbreak occurs in the Motaba

River Valley of Zaire, it's called the Motaba virus. We learn that the disease is invariably fatal and that no one infected lives more than two days. "This is the scariest son of a bitch I've ever seen," says Hoffman.

In the next scene, a monkey that had escaped from the village is trapped and put on a ship headed for Seattle.

Back at Fort Detrick, scientists have isolated the virus, which by electron microscopy looks exactly like the Ebola virus. We learn that the virus is so deadly that it can reproduce itself in cells in only one hour. (Unlike bacteria, viruses can grow only inside cells, which usually takes two to three days. One hour would be impossible.)

The monkey that was captured in Zaire ends up in a Biotest Animal Holding Facility in San Jose, California, where it is promptly stolen by Patrick Dempsey and sold to the owner of Rudy's Pet Shop in Cedar Creek. Both Dempsey and the pet-store owner later die from the disease—but not before Dempsey releases the monkey into the forest surrounding Cedar Creek.

People in Cedar Creek start dying from the Motaba virus. The city is put under military control; no one can enter or leave. Then something unusual happens (not that all of this hasn't been unusual enough). The Motaba virus infects a man who has been hospitalized for weeks following a car accident—a man who was never exposed to the monkey. How did he catch the virus? Hoffman realizes that the virus has mutated, now easily spreading through hospital air ducts (this is the gain-of-function part).

In an effort to control the outbreak, the military decides to bomb Cedar Creek, killing all twenty-six hundred of its residents. The bombing mission is called Operation Clean Sweep. When Dustin Hoffman finds out about the proposed bombing, he knows that the race is on to find a lifesaving antiserum that will save Cedar Creek's citizens, which now includes his ex-wife (they've recently reconciled).

Hoffman believes that if he can find the monkey that started the outbreak, he can make the antisera he needs. Much of the movie is now spent trying to find that darn monkey, including one dramatic helicopter chase scene in which the military tries to shoot Hoffman out of the sky. (Hoffman doesn't really need the monkey. He's already isolated and grown the virus. All he needs now is to inoculate the virus into one of any number of animals and make his antisera. Forget about the monkey.) Once the monkey is in hand, Hoffman instructs Cuba Gooding Jr. to take it back to the lab, determine which antibodies are neutralizing the mutant virus, synthesize those antibodies, and make several liters of life-saving antisera. Assuming everything goes well, Hoffman's task should take about a year. Cuba Gooding Jr. does it in a little less than a minute. (Now I understand why people are angry that we still don't have an AIDS vaccine.) Then, in one final if-only-this-were-true-in-real-life flourish, after receiving the antiserum, Rene Russo improves in thirty seconds.

Although it is no doubt out of their comfort zone, scientists should try to do what Ian Lipkin did in *Contagion*: get involved in movies about science, whether fiction or nonfiction. It's worth it. For example, starting in the 1950s, formal relationships between NASA and the American Medical Association and the movie and television industry have been shaping our perceptions of astronauts and doctors. Indeed, NASA, by consulting on movies like *Apollo 13* (1995), *Armageddon* (1998), *Mission to Mars* (2000), and *Space Cowboys* (2000), was able to maintain its positive image despite the Challenger disaster of 1986.

Other examples abound. The participation of the National Severe Storms Laboratory in *Twister* (1996), the United States Geological Survey in *Dante's Peak* (1997), and Federal Express in *Cast Away* (2000) led to better perceptions of those organizations. One group that could clearly benefit from a better working relationship with the film industry is vaccine makers. No group, it seems, has a worse reputation than pharmaceutical companies. Indeed, in the movie *The Constant Gardener* (2005), a pharmaceutical company (presumably using its black ops division), murdered people who had discovered a deadly side effect from its new drug. This plot was believable to most viewers.

The movie *The Lost World: Jurassic Park* (1997) provides one final inducement for scientists to serve as consultants. The paleontologist Jack Horner, who consulted on the film, was in the midst of an argument with a colleague about a particular concept. Horner later noted that one of the actors who gets eaten by a *T. rex* in the film bore "a striking resemblance to a guy who argues with me a lot."

"If you are ever arguing with someone," warns Horner, "don't let them be the advisor on a movie."

CHAPTER 8

The Emperor's New Clothes

Narcissus does not fall in love with his reflection because it is beautiful, but because it is his. If it were his beauty that enthralled him, he would be set free in a few years by its fading.

—W. II. AUDEN

Charisma sells.

When trying to communicate science and health information to the public, celebrities can be enormously helpful—or harmful. For example, in the early 1950s, Arthur Godfrey, a beloved radio and television entertainer, promoted cigarette smoking. Godfrey created the slogan "Buy 'em by the carton." Known affectionately as the "Old Redhead," Arthur Godfrey was an avuncular, big-hearted figure. Even though studies had already shown that cigarette smoking caused lung cancer—and that the more people smoked, the greater the risk—Godfrey was influential. People trusted him. If Arthur Godfrey said cigarette smoking couldn't hurt you, then scientists were probably wrong. Godfrey later died from emphysema, a complication of the radiation therapy he had received to treat his lung cancer.

In the early 2000s, Charlton Heston, a popular actor, lobbied against gun control, even though countries with stricter gun laws had fewer gun-related crimes. Heston was best known for his roles as Moses in *The Ten Commandments* (1956), Ben-Hur in *Ben-Hur* (1959), and George Taylor in *Planet of the Apes* (1968). Heston was also the five-term president of the National Rifle Association (NRA). On May 20, 2000, while addressing an annual meeting of the NRA, Heston said, "For the next six months, [presidential hopeful] Al Gore is going to smear you as the enemy. He will slander you as gun-toting, knuckle-dragging, bloodthirsty maniacs who stand in the way of a safer America. Will you remain silent? I will not remain silent. If we are going to stop this, then it is vital to every law-abiding gun owner in America to register to vote and show up at the polls on Election Day." Heston picked up a rifle. "I want to say these fighting words for everyone within the sound of my voice to hear and to heed, and especially for you, Mr. Gore: 'From my cold, dead hands!'" Charlton Heston was a powerful, dynamic figure—when he talked, people heard the voice of God. "From my cold dead hands" became a bumper sticker and a rallying cry for the NRA.

On June 24, 2005, during an interview with Matt Lauer on NBC's *Today*, Tom Cruise, another popular actor, became the national spokesperson for Scientology. He did it by denying the validity of Brooke Shields's postpartum depression, saying, "Psychiatry is a pseudoscience."

"The thing that I'm saying about Brooke is that there's misinformation, okay?" said Cruise. "There is no such thing as a chemical imbalance in a body."

Cruise, who had starred in *Top Gun* (1986), *Rain Man* (1988), *A Few Good Men* (1992), *Mission Impossible* (1996), and *War of the Worlds* (2005) was a trusted voice. "You don't know the history of psychiatry," he told Lauer. "I do."

Cruise knew what Scientologists had taught him. Specifically, that seventy-five million years ago, Xenu, the tyrant ruler of the Galactic Confederacy, had brought billions of people to Earth in a spacecraft, stacked them around volcanoes, and set off hydrogen bombs. The explosion released spirits called "thetans," which caused psychological harm. To the Scientologist, psychotherapy centers on neutralizing these thetans.

Vaccines have also suffered their share of misguided celebrities. One man, however, who is neither a movie actor nor a television star, has risen above the rest: Andrew Wakefield, a handsome, square-jawed, well-spoken physician-scientist with a British accent and a firm resolve—a celebrity scientist. "He is very charming and very convincing," wrote Matt Carey, the curator of the blog *Left Brain/Right Brain* and the father of a child with autism. "If this were Hollywood central casting and they cast him as himself, people would say, 'Hollywood has gone over the top.'" No one has done more to promote the anti-vaccine movement than Andrew Wakefield. And, ironically, no one has done more to destroy it.

To understand how Andrew Wakefield became an international spokesperson for anti-vaccine activists, we need to understand where he came from, what he proposes, and why he persists.

In 1981, Wakefield, the son of two doctors, received his medical degree from St. Mary's Hospital Medical School, of Imperial College London. He chose a career in gastrointestinal surgery. In 1985, he was made a fellow of the Royal Academy of Surgeons. The following year, he was awarded the coveted Wellcome Trust traveling fellowship to study intestinal transplantation at the University of Toronto in Canada. When he returned to London, he was hired by the Royal Free Hospital, one of the most prestigious hospitals in the United Kingdom. Wakefield studied inflammatory bowel disease, becoming one of the first to describe specific blood flow abnormalities in the intestines. Richard Horton was at the Royal Free when Wakefield made his discovery. "I was in a different department to Wakefield," recalled Horton, "but close enough to see the sensation it caused. Research in the Royal Free's Academic Department of Medicine was largely moribund at the time I was there. [But] Wakefield brought a sudden sense of excitement. He was a committed, engaging, and charismatic clinician and scientist. The department felt alive again. He asked big questions about diseases, and his ambition brought quick and impressive results." In the early 1990s, Andrew Wakefield was at the top of his game.

In 1995, Rosemary Kessick, the mother of a child with autism, approached Andrew Wakefield. Kessick, who was convinced that the MMR vaccine had caused her son's autism and intestinal symptoms, had come to the right doctor. For the next three years, Wakefield devoted himself to studying a group of children with similar stories.

In 1998, Wakefield published his findings in *The Lancet*, one of the oldest, most respected medical journals in the world. Wakefield described twelve children—eight of whom had autism—whose parents believed that they had lost language, communication, and social skills within weeks of receiving the MMR vaccine. Wakefield found that these children also had intestinal inflammation. In his paper, Wakefield didn't explain how the intestinal inflammation had caused autism. Rather, he waited until the next day during a press conference.

Wakefield connected the dots. He said that because the measles vaccine had been given in combination with the mumps and rubella vaccines, the children's immune systems had been overwhelmed. Because their immune systems had been weakened, the measles vaccine virus could now travel to the intestine unchecked, reproduce, and cause intestinal inflammation. With their intestines inflamed, the children could no longer rid themselves of toxins in food. These toxins then entered the bloodstream, traveled to the brain, and caused autism. Wakefield then said something that alienated him from doctors, scientists, and public health officials in the United Kingdom and around the world. He said that children should no longer receive the MMR vaccine. Rather, the measles, mumps, and rubella vaccines should be separated. Because separate vaccines weren't readily available, Wakefield had, for all practical purposes, advised parents not to protect their children against these three diseases.

Among many parents of children with autism, Andrew Wakefield was an immediate hero. At last, a doctor who listened, a doctor who cared, a doctor who not only knew what caused autism but, by restricting children's diets, knew how to treat it. During the next few years, more than fifteen hundred articles about the MMR vaccine, autism, and Wakefield appeared in newspapers and magazines across the globe. Andrew Wakefield had become a media sensation.

In 2003, the British Broadcasting Corporation (BBC) aired a docudrama about Wakefield's life and work. Titled *Hear the Silence*, it starred two popular British actors: Juliet Stevenson as Christine, a mother searching for the cause of her son Nicky's autism, and Hugh Bonneville as Dr. Andrew Wakefield.

In *Hear the Silence*, Christine is shuffled from one medical specialist to another, each saying that nothing can be done. But Christine believes that something *can* be done. She just has to find out what caused her son's descent into autism.

Poring over old photographs, Christine notices that Nicky had stopped smiling and talking soon after he had received the MMR vaccine. Eventually, she walks into the office of Dr. Wakefield, the first doctor who takes her concerns seriously. "You believe me?" asks Christine, amazed.

"Of course," says Wakefield. "Why wouldn't I believe you?"

Exhausted, overwhelmed from fighting a medical system that had stubbornly denied what she believed to be true, Christine bursts into tears. Wakefield smiles at her beatifically.

Committed to proving Christine's theory, Wakefield performs colonoscopies on a series of children who, like Nicky, developed autism after receiving the MMR vaccine. In *Hear the Silence*, Wakefield, over a series of dramatic animation sequences, explains that the measles virus usually enters the body through the nose and throat, not the muscles (ominous-looking measles vaccine viruses are injected into a muscle). For this reason, the vaccine virus takes up residence in the gut, where it causes inflammation (a red stain spreads across the intestinal surface). Because the MMR vaccine contains three viruses, children's immune systems are overwhelmed, causing chronic ear infections and bronchitis (a child, haggard, with puffy eyes and a fever, coughs). Mothers feed the fever, inadvertently giving children "toxins" like gluten in bread or casein in milk, which can't be digested by a damaged intestine (a child eats a cheese pizza). Viewers of *Hear the Silence* are then told that gluten and casein are morphine-like toxins that travel to the brain and cause autism (a black stain spreads across the brain).

"It was an interior *Jaws*," wrote David Aaronovitch for London's *The Observer*, "The MMR vaccine is coming to get our kids."

In the next scene, we find that Andrew Wakefield was right. After Christine restricts his diet, Nicky calms down and begins to talk.

Back in the laboratory, Wakefield is looking through a microscope. A senior physician hands him a letter. Wakefield opens it, then slowly exhales. "*The Lancet* has agreed to publish our paper," he says. Colleagues cheer, patting each other on the back, shaking hands, all about to publish the most important finding of their scientific careers—all unaware of the firestorm they are about to unleash.

Hear the Silence ends on a note of hope. Gathered in a park, a group of parents praise Andrew Wakefield, who sits Christlike among them. "It's hard to be a prophet in your own land," Christine tells Wakefield. Then, speaking to the other parents, she says, "I used to think I was alone. But there are more and more of us every week, and that's the biggest tragedy of all. We can't stop now. We can never stop."

Back home, Wakefield opens a letter, then looks at his wife. "I've been asked to speak in front of Congress," he says.

In the final scene, Wakefield stands in front of the United States Capitol building, takes a deep breath, and walks forward.

With a *Lancet* publication, a made-for-television biopic, and the support of thousands of parents of children with autism, Andrew Wakefield had become the rarest of rarities: a famous scientist. A closer look, however, shows how Wakefield's paper was the scientific equivalent of the Emperor's New Clothes. There was simply nothing there.

To explain how the MMR vaccine caused autism, Andrew Wakefield had proposed a series of implausible events—none of which were supported in his *Lancet* paper.

First, Wakefield claimed that by combining the measles, mumps, and rubella vaccines into a single shot, children's immune systems had been overwhelmed. In 1998, when Wakefield published his paper, skin and blood tests to determine immune dysfunction were readily available. Wakefield either didn't do these tests or didn't report the results.

Second, Wakefield claimed that because the MMR vaccine had weakened children's immune systems, the measles vaccine virus had traveled to and damaged the intestines. Although Wakefield had performed intestinal biopsies, he didn't test those samples for measles vaccine virus proteins. "Virological studies are underway that may help to resolve this issue," he wrote. Again, if Wakefield was going to claim that the measles vaccine virus was destroying intestinal cells, the reviewers should have required him to prove it.

In 2008, ten years after Wakefield published his *Lancet* paper, Mady Hornig and colleagues at Columbia University did the studies that Wakefield should have done. Examining intestinal biopsies from children with and without autism, *including those from Wakefield's Lancet paper*, Hornig found no differences in the presence of the measles vaccine virus between the two groups. Three independent laboratories later confirmed these findings.

Hornig's results shouldn't have been surprising. In Wakefield's report, one child's symptoms of autism had begun only *one day* after immunization. This simply isn't enough time for the measles vaccine virus to reproduce (that takes two to three days), damage the intestine, cause intestinal inflammation, and allow for the entrance of toxic proteins into the bloodstream that travel to the brain, damage the brain, and cause autism. The fact that one child in his report had autism within a day of receiving the MMR vaccine should have caused Wakefield to abandon, or at least question, his hypothesis.

Third, Wakefield claimed that children with autism were unable to rid themselves of gluten and casein. To prove it, all he had to do was test the blood and spinal fluids that he had already collected. Since Wakefield first proposed his theory, studies have shown that children with autism whose diets are restricted don't fare better than those whose diets aren't restricted. The only difference is that children who are deprived of calcium and vitamin D in dairy products are more likely to develop a dangerous thinning of their bones called osteoporosis. Wakefield's restrictive diet worked in *Hear the Silence*, but not in real life.

Finally, when Andrew Wakefield published his paper in 1998, about one of every two thousand children in England was diagnosed with autism. Also at that time, about 90 percent of children in England were receiving the MMR vaccine. Given these percentages, and given that the MMR vaccine was administered at the beginning of the second year of life—when many children are diagnosed with autism—one would have expected that statistically every year, about three hundred children would have developed autism within one month of receiving the MMR vaccine. The only way that this wouldn't have happened would have been if the MMR vaccine prevented autism. Which it doesn't. It only prevents measles, mumps, and rubella infections. In his *Lancet* paper, Wakefield issued the following challenge: "If there is a causal link between [the] measles, mumps, and rubella vaccine and this syndrome, a rising incidence might be anticipated after the introduction of this vaccine in the UK in 1988." Wakefield was proposing the obvious: Let's look at the incidence of autism before and after the introduction of the MMR vaccine. Since he issued his challenge, seventeen studies performed in Australia, Denmark, England, Finland, Japan, Poland, and the United States have examined the incidence of autism in hundreds of thousands of children who either did or didn't receive the MMR vaccine. All of these studies found the same thing: The MMR vaccine hadn't caused autism.

In December 2001, unwilling to support either Andrew Wakefield or his research, the Royal Free Hospital asked him to resign. "I have been asked to go because my research results are unpopular," he said. Wakefield's choice of the word *unpopular* was a curious one. He wasn't fired because he was unpopular. He was fired because he was dangerous. Three years earlier, with no data to support his theory, Wakefield had scared the hell out of British parents. Massive outbreaks of measles, hundreds of hospitalizations, and four deaths were the result. After his firing, Wakefield said, "I did not wish to leave, but I have agreed to stand down in the hope that my going will take the political pressure off my colleagues. . . . I have not done anything wrong. Losing a hospital teaching job does not do much for my [résumé], but there are bigger issues at stake. What now matters most is what happens to these children." Despite his firing, many parents of children with autism stood by him.

In February 2004, Andrew Wakefield left England and traveled to Melbourne, Florida, where he became the director of research at the International Child Development Resource Center. Two men, Drs. Jeffrey Bradstreet and Jerry Kartzinel, had founded the Center, calling it the home of the "Good News Doctor." The good news was that Bradstreet and Kartzinel believed, like Wakefield, that the MMR vaccine caused autism and that they, too, could treat it. (Kartzinel would later publish a book with Jenny McCarthy titled *Healing and Preventing Autism: A Complete Guide*.) Bradstreet and Kartzinel, along with a group of doctors who were part of a group called Defeat Autism Now!, claimed to have successfully treated hundreds of children with autism. Some of the treatments used by at least one of these clinicians included dietary supplements, restrictive diets, antifungal and antiviral drugs, coffee enemas, intravenous medicines that bound heavy metals, a plant from the ginger family, digestive enzymes, electrical and magnetic stimulation of the brain, hookworms, whipworms, hyperbaric oxygen chambers, lymphatic drainage massage, probiotics, intravenous immunoglobulins, naltrexone suspended in emu oil, and chiropractic manipulations. Wakefield fit right in. But not for long.

One year later, Andrew Wakefield found a better gig. He left Florida and traveled to Austin, Texas, to become the executive director of Thoughtful House, which advertised, "If your child is diagnosed with autism, the next thing to know is that autism is treatable."

On January 28, 2010, Andrew Wakefield suffered another setback. Spurred by the revelations of an investigative journalist named Brian Deer, officials from the General Medical Council (GMC) in London took a closer look at Wakefield's research. Deer had found that "not one of the twelve cases reported in the 1998 *Lancet* paper was free of misrepresentation or undisclosed alteration, and that in no single case could the medical records be fully reconciled with the descriptions, diagnoses, or histories published in the journal." According to Deer, Andrew Wakefield had falsified medical records. Deer also exposed the source of funding for Wakefield's study. When the GMC finished its investigation, it banned Andrew Wakefield from practicing medicine in England. With the help of Brian Deer, the GMC found that Wakefield had (1) received £435,643 ($630,000) in fees plus £3,910 ($5,660) in expenses through personal injury lawyers in the midst of suing pharmaceutical companies for damages caused by the MMR vaccine; (2) misled the *Lancet*, as well as patients who had participated in the study regarding the purpose of his research, creating the impression that children had been chosen randomly when in fact several were part of a lawsuit, essentially laundering legal claims through a medical journal; (3) performed unnecessary and invasive procedures like colonoscopies and spinal taps on children without getting approval from his hospital's ethics committee; (4) taken blood from children who had attended his young son's birthday party, paying each child £5 ($7.20), a practice considered to be unethical in the United Kingdom, and later joking about how several had vomited.

The GMC said that Wakefield had "callously disregarded the pain and distress young children might suffer and behaved in a way that brought the profession into disrepute." Further, the Council stated that Wakefield had been "dishonest and irresponsible." In the end, the GMC found Wakefield guilty of more than thirty charges, including four counts of dishonesty and twelve counts of abusing developmentally challenged children.

GMC officials could have banned Wakefield from practicing medicine for one year, but they didn't. Instead, they banned him from ever practicing medicine in England again.

After the GMC revoked his medical license, Andrew Wakefield, appearing on NBC's *Today*, described the ban as "a little bump in the road" and said that he was committed to continuing his research. "There are millions of children out there suffering," he said. "And the fact is that the vaccines cause autism. These parents are not going away. The children are not going away. And I am most certainly not going away."

Again, parents of children with autism stood by his side. "The GMC can say whatever they want to say for the rest of their existence," said Rebecca Estepp, the head of a prominent anti-vaccine organization and the mother of a child with autism. "But I know that my son got better because of Andrew Wakefield."

On February 2, 2010, twelve years after Wakefield had published his paper, the *Lancet* retracted it. Technically, it no longer exists. The official retraction was supported by ten of Wakefield's twelve coauthors. Papers aren't typically retracted because the science is flawed; they're retracted because the science is fraudulent. Richard Horton, the editor-in-chief of the *Lancet*, said, "It's the most appalling catalogue and litany of some of the most terrible behavior in any research, and it is therefore very clear that it has to be retracted." Later, Horton lamented, "The statements in the paper were utterly false. I feel I was deceived."

No one, however, took Andrew Wakefield to task more than Fiona Godlee, the editor-in-chief of the *British Medical Journal*. "Who perpetrated this fraud?" she wrote. "There is no doubt that it was Wakefield. Is it possible that he was wrong, but not dishonest; that he was so incompetent that he was unable to fairly describe the project, or to report even one of the twelve children's cases accurately? No. A great deal of thought and effort must have gone into drafting the paper to achieve the results he wanted: The discrepancies all led in one direction;

misreporting was gross." Godlee also challenged Wakefield to do what he said he was going to do: exonerate himself with studies proving that his theory was right. "Furthermore," wrote Godlee, "Wakefield has been given ample opportunity either to replicate the paper's findings, or to say he was mistaken. He has declined to do either. He refused to join ten of his coauthors in retracting the paper's interpretation in 2004, and has repeatedly denied doing anything wrong at all. Instead, although now disgraced and stripped of his clinical and academic credentials, he continues to push his views."

Andrew Wakefield's curtain call on the national stage occurred on January 5, 2011, on *Anderson Cooper 360°*. During this interview, which followed Brian Deer's stunning exposé and the *British Medical Journal*'s description of Wakefield's *Lancet* paper as an "elaborate fraud," Cooper was unrelenting. "The editors of the BMJ are saying that Wakefield's study wasn't just bad science," says Cooper, "but an elaborate fraud that's done long-lasting damage to the public's health. How do you respond?"

"I've had to put up with [Brian Deer's] false allegations for many, many years," says Wakefield. "He's a hit man. He's been brought in to take me down because they are very, very concerned about the adverse reactions to vaccines that are occurring in children."

"Who are 'they'?" asks Cooper. "Who is he a hit man for? Are you saying that this is some sort of conspiracy against you? Is that your argument?"

"*Conspiracy* is your word," argues Wakefield. "What it is is a pragmatic attempt to crush any investigation into valid vaccine safety concerns, not just my concerns." (Wakefield now holds up his book *Callous Disregard* for the camera.)

"Sir, I'm not here to let you pitch your book," says Cooper.

Wakefield isn't dissuaded: "If you read the record that I have set out in the book, you will see the truth."

"But, sir, if you're lying, then your book is also a lie," counters Cooper.

"The book is not a lie; the study is not a lie," pleads Wakefield.

Cooper refuses to back down. "If your research was valid, why have ten of your coauthors all retracted the paper's interpretation back in 2004?"

"Because I'm afraid that the pressure has been put on them to do so," says Wakefield. "And people get very frightened. You're dealing with some very powerful interests here."

Cooper then interviews Brian Deer, who explains that he has never received money from pharmaceutical companies or their representatives and that he has recently won an award for a series of articles attacking the pharmaceutical industry. Deer further explains that he has personally talked to four of the families in Wakefield's study and that a journalist working on his behalf had talked to two others. He explains that he has visited one California family who said that the description of their son in Wakefield's paper was "not true."

After the Anderson Cooper interview, Wakefield's television appearances were limited to programs like *The Alex Jones Show*. Jones, like Wakefield, is a conspiracy theorist, believing that the United States government was behind the 1995 Oklahoma City bombing and the September 11, 2001, attacks, that the moon landing was filmed on a Hollywood sound stage, and that the tragic shootings at Sandy Hook Elementary School in Newtown, Connecticut, in 2012, were a hoax.

After his rejection by mainstream media, Andrew Wakefield slipped further into the abyss.

In February 2016, Wakefield climbed aboard a Ruby Princess cruise ship headed for Mexico. He'd been asked to participate in a symposium. The nature of the symposium, the backgrounds of his fellow speakers, and the interests of the attendees, show just how far Andrew Wakefield's star had fallen.

The Ruby Princess's floating symposium also featured Robert O. Young, who believed that all cancers were caused by excess acidity and that his "alkaline diet" was the cure; Nick Begich, who believed that the government used weather and AIDS drugs as a form of mind control; Sean David Morton, who used psychic powers he had learned from Nepalese monks to predict the stock market and avoid taxes (Morton and his wife were later arrested for tax evasion); Jeffrey Smith, an anti-GMO activist, who taught "yogic flying"; Dannion Brinkley, who believed that he had died and gone to heaven three times (death, he argued, was no big deal); Leonard Horowitz, who believed that the hepatitis B vaccine was the origin of AIDS; and others, who talked about giant lizards called "Reptilians," dark-energy beings called "Dracos," and an "incarnate fairy" who had successfully moved the prime meridian. They also talked about how the government was injecting something strange into the chicken served at Popeyes and Church's Chicken restaurants. As described by Anna Merlan, writing for the blog *Jezebel*, throughout the cruise, power cords went missing, projectors broke, and slides froze. Merlan called it the "Conspira-Sea Cruise": America's largest floating gathering of conspiracy theorists.

When it was Wakefield's turn to speak, he urged reporters to watch his presentation. Unfortunately, most had already committed to a talk about "wishing machines." Snubbed, Wakefield stood in front of about a hundred retirees in a dining room on a ship in the middle of the ocean and talked about how the federal government and pharmaceutical companies had conspired to suppress the truth about the MMR vaccine. "I've sat in this field now for twenty years," he said, "and nothing has persuaded me that the science is wrong." When he was finished, Wakefield opened the floor for questions. One attendee asked him to comment on the "twelve alternative doctors last year" who had been murdered by the government—murders that had been made to look like suicides.

At the end of the cruise, Helen Sewell, an astrologer, praised the attendees and speakers. "You're all Promethean spirits," she enthused. "You've all awakened, and you're shining your Uranian light into Pluto's underworld."

Thirty years earlier, Andrew Wakefield had been a rising star in the academic world. Now he was on a roster of speakers who were talking about elves, fairies, and intergalactic wars. What happened? Why had Wakefield allowed himself to fall so far? Surely, he wasn't the first scientist to publish a paper that was wrong. And he wasn't the first scientist to publish a paper that was retracted. But few scientists have ever held on to a theory that had been so thoroughly discredited for so long. Why?

I don't really know the answer to this. But I'm going to propose three possible theories.

The Medical-Establishment-Doesn't-Care-About-You-but-I-Do Theory

Medicine has limits. For some disorders, like autism, despite hundreds of millions of dollars spent trying to unravel its secrets, scientists have yet to identify a single cause or cure. Although behavioral therapies can help, for the most part, medicine has little to offer. From the standpoint of parents financially and emotionally burdened by autism, doctors appear to be throwing up their hands and saying, "There's really nothing we can do" (which is exactly what one of the doctors said in *Hear the Silence*). On the other hand, doctors who offer answers appear to be the only ones who really care.

When Andrew Wakefield claimed that restrictive diets could cure autism, he created a devoted following. Simple solutions, even when they're wrong, are enormously compelling. After his *Lancet* paper was retracted and he had lost his license to practice medicine, many parents still stood behind him. "He was the first doctor who really listened to us," said Isabella Thomas. "This man was my savior." For some physicians, this adoration can be quite seductive.

It's hard to resist being revered.

The I've-Fallen-In-Love-with-My-Own-Reflection Theory

Andrew Wakefield believes that he and he alone has seen the light. "I have no interest in what my colleagues think of me," he has said. One could reasonably argue that Wakefield's grandiosity, his exaggerated sense of self-importance, his fantasies of brilliance, his sense of entitlement, his need for constant admiration, and his arrogance meet the definition for narcissism.

Supporting the case for Wakefield's narcissism is the book he wrote in 2010 titled *Callous Disregard: Autism and Vaccines— The Truth Behind a Tragedy*. *Callous Disregard*, which refers to a phrase used by the GMC when it withdrew Wakefield's license to practice medicine in England, is a painstaking, point-by-point refutation of almost every real or perceived slight Andrew Wakefield has ever suffered. In *Callous Disregard*, Wakefield argues that

- Every aspect of his *Lancet* paper was not only correct, but groundbreaking.
- All thirty charges of misconduct, misrepresentation, and unethical behavior levied against him by the GMC were ill founded and incorrect. It's interesting to note that while Wakefield defended himself in his book, he did little to defend himself in front of the GMC. On May 24, 2010, the GMC panel reported, "On behalf of Dr. Wakefield, no evidence has been adduced, and no arguments or pleas in mitigation have been addressed to the Panel." Rather than defend himself at his licensing hearing, Wakefield chose to take his case to the people.
- Practically everyone involved in discrediting him has either been in the pocket of a corrupt government or a greedy pharmaceutical company. Andrew Wakefield stands alone in his righteousness.
- The money he received from personal-injury lawyers in the midst of suing pharmaceutical companies was acquired openly, honestly, and above board (even though his colleagues apparently didn't know about it).
- Subjecting children to invasive procedures like colonoscopies and spinal taps was not only the right thing to do, but a necessary thing to do.

- Everyone who has ever said anything negative about him is either ignorant or deluded, including the dean of the Royal Free Hospital School of Medicine and professors and clinicians from many other academic institutions around the world.

Wakefield was particularly enraged when he found a biology exam that had included his *Lancet* paper as an example of bad science, the instructor using it as a negative lesson. He compared this to the Nazis poisoning young German minds with dangerous propaganda. Referring to Julius Streicher, a prominent member of the Nazi Party who had published an anti-Semitic book for children titled *Der Giftpilz* (*The Poisonous Mushroom*), Wakefield wrote, "This is such utter garbage that one wonders whether Julius Streicher actually survived the hangman's noose in Nuremberg with little more than a whiplash, only to return as a graying biology teacher needing to make a little money on the side." (Always the Nazi analogy.)

Further support for the I've-fallen-in-love-with-my-own-reflection theory comes from a documentary film about Andrew Wakefield titled *The Pathological Optimist*. Released in September 2017, this film showed Wakefield's darker side. Throughout the movie Wakefield travels from venue to venue to speak to groups of parents of children with autism, at one point likening himself to Nelson Mandela. But he isn't trying to raise money to study the cause or causes of autism. Nor is he trying to raise money to provide support services for these children. Rather, he is taking advantage of a vulnerable group of parents to financially support his lawsuits against Brian Deer, the investigative reporter who had exposed his financial improprieties, and Fiona Godlee, the editor-in-chief of the *British Medical Journal* who had called him a fraud. In *The Pathological Optimist*, Wakefield is portrayed as someone who is interested in himself and his reputation only. The children appear to be an afterthought. (Wakefield lost both lawsuits.)

The Walking-Along-the-Sea-of-Galilee Theory

This theory is a variant of the narcissism theory.

In December 2000, two years after he published his *Lancet* paper, Andrew Wakefield published a paper titled "Measles, Mumps, Rubella Vaccine: Through a Glass Darkly." "Through a glass darkly" is a phrase from St. Paul's letter to the Corinthians. The full quote is "For now we see through a glass darkly; but then face to face; now I know in part; but then shall I know even as also I am known" (1 Corinthians 13:12). The passage means that although we might not understand something now, everything will be revealed when we meet our creator. A few years later, Wakefield was speaking to a group of followers at Trinity Christian College in Chicago. Ken Reibel, who runs the blog *Autism News Beat*, attended the event. "Wakefield shared the stage with a large wooden cross," wrote Reibel. "The stained glass window over his head read, 'He is before all things and in Christ all things hold together.' Wakefield warmed up the audience by asking how many had an autistic child. Then he asked how many thought that vaccines were the cause. Then he asked, 'How many here tonight are representing a pharmaceutical company? There's always one or two. I know who you are. That's OK. You are welcome.'" Wakefield had apparently forgiven them for their sins.

Recently Wakefield said that children treated by chiropractors "will inherit the earth." "They are extraordinarily healthy because of the lifestyle they live and the fact that so many of them are unvaccinated," he argued. Wakefield has said that he would give his own life so that children with autism would no longer have to suffer. For a man who doesn't profess to be particularly religious, he makes a lot of religious references. Psychiatrists informally refer to this collection of behaviors as the messiah complex.

While you might think that this would turn off his followers, the opposite appears to be true. J. B. Handley, a prominent anti-vaccine activist, once referred to Wakefield as being a combination of Jesus Christ and Nelson Mandela.

Today, 10 to 20 percent of American parents choose to delay, separate, space out, or withhold vaccines for their children; 2 percent don't give them any vaccines at all. Many of these parents remain concerned that vaccines might cause autism. As a consequence, outbreaks of measles and mumps are sweeping across the nation. In 2014, a measles outbreak involving seven hundred people was larger than any seen in more than two decades. In 2015, a measles epidemic starting in Disneyland engulfed twenty-five states. In 2016, more than six thousand children suffered from mumps, a disease that had caused fewer than two hundred cases a few years earlier. In 2017, a measles epidemic broke out in a Somali community in Hennepin County, Minnesota. Several years earlier, Andrew Wakefield had held special meetings with Somali parents to warn them of the dangers of the MMR vaccine. Before Wakefield's visits, 92 percent of Somali children had been vaccinated; after his visits, that number dropped to 42 percent. Later, when a *Washington Post* reporter asked him whether he felt responsible for the outbreak, Wakefield said, "I don't feel responsible at all."

CHAPTER 9

Judgment Day

The Democrats are the party that says government will make you smarter, taller, richer, and remove the crabgrass on your lawn. Republicans are the party that says government doesn't work, and then they get elected and prove it.

—P. J. O'Rourke, *Parliament of Whores: A Lone Humorist Attempts to Explain the Entire U.S. Government*

Like celebrities, politicians can also use their considerable platforms to inform or misinform the public about science.

At the end of the biopic *Hear the Silence*, Andrew Wakefield, played by the actor Hugh Bonneville, receives a letter asking him to testify before the United States Congress. In the final scene, Wakefield stands on the steps of the Capitol building, then walks confidently forward. Wakefield's letter had come from Representative Dan Burton, a Republican from Indiana, who had subpoenaed him to appear before the powerful House Committee on Government Reform. The hearing took place on April 6, 2000. I had also been asked to testify.

The title of the hearing was "Autism: Present Challenges and Future Needs—Why the Increased Rates?" I didn't understand why I had been subpoenaed. I wasn't an autism expert, so I couldn't talk about the cause or causes of autism. And I wasn't an epidemiologist, so I couldn't talk about "why the increased rates?" I assumed, given that I was a virologist and immunologist, that Dan Burton wanted me to talk about whether the MMR vaccine could cause autism (which tells you everything you need to know about how naïve I am). I would be given ten minutes to read from a prepared statement and then, presumably, answer questions. For the next two weeks I worked hard on my speech. The real reason that Burton had asked me to testify wouldn't become clear until later.

At 10:37 a.m. in room 2154 of the Rayburn House Office Building, Dan Burton called the hearing to order. He sat on a high dais, flanked on either side by his fellow committee members. The room was packed; the atmosphere tense, expectant, and foreboding. Although I understood the seriousness of the meeting, I felt like I was in a Monty Python skit about Judgment Day.

It didn't take long to see where the hearing was headed. "This morning we are here to talk about autism," said Burton. "What used to be considered a rare disorder has become a near epidemic. We have received hundreds of letters from parents across the country. [But] I do not have to read a letter to experience the kind of heartbreak that is in these letters. I see it in my own family." Burton showed a slide of photos of his two grandchildren.

The one on the left is my granddaughter, who almost died after receiving a hepatitis B shot. Within a short period of time, she quit breathing, and they had to rush her to the hospital. My grandson, Christian, whom you see there with his head on [his mother's] shoulder, according to the doctors was going to be about six foot ten. We anticipated having him support the family by being an NBA star. But unfortunately, after receiving nine shots in one day, the MMR and the DTaP shot, and the hepatitis B, within a very short period of time, he quit speaking, ran around banging his head against the wall, screaming, hollering, waving his hands, and became a totally different child. We found out that he was autistic. He was born healthy. He was beautiful and tall. He was outgoing and talkative. He enjoyed company and going places. Then he had those shots, and our lives and his life changed.

Burton stopped, fighting back tears.

At this point I realized that the subtitle of the hearing, "Why the Increased Rates?," had been a ruse. We weren't here to find out why the rates of autism had increased. Dan Burton already believed he knew why. And he had invited a series of panelists—most prominently Andrew Wakefield—to prove his point.

After Burton finished his opening statement, Representative Henry Waxman, a Democrat from California, gave his. "I believe that we need to increase our efforts to understand the causes of autism," he said. "But in this process we must not get ahead of the science or raise false alarms. [I]n medicine the best answers come from research that can withstand the rigors of the scientific method. These standards have been developed in order to find the truth." Where Burton had been emotional and anecdotal, Waxman was rigorous and thoughtful. I started to relax a little bit. At least one member of Congress would be on my side.

The first panel consisted of parents of children with autism. All, save one, believed that vaccines had been the cause. The parent who didn't was Wayne Dankner, a pediatric infectious diseases specialist from San Diego, who warned, "I have seen no sound evidence linking autism to MMR or any other vaccine."

Burton grilled Dankner, trying to find some evidence that Dankner's daughter had worsened following a vaccine, any vaccine. But Dankner refused to budge. Burton tried to salvage the moment by performing an impromptu epidemiological ministudy. He asked each of the parents who had already testified to pinpoint exactly when after vaccination their children had developed autism. Satisfied that his theory was still intact, he moved on.

The next group to testify was the scientists. Andrew Wakefield was first. Speaking to members of Congress as if they were molecular biologists, Wakefield talked about measles virus proteins, follicular dendritic cells, ileocolonic lymphonodular hyperplasia, crypt abscesses, common recall antigens, and molecular-amplification technology. Burton nodded enthusiastically throughout Wakefield's presentation, no doubt having little to no idea what he was talking about. Next up was John O'Leary, a pathologist from Ireland and a colleague of Wakefield's who, like Wakefield, buried the committee in jargon. O'Leary talked about TaqMan real-time quantitative polymerase-chain reactions, RNA inhibition assays, low-copy viral gene detections, fusion proteins, nucleocapsid genes, and black signals, concluding that his detection of measles virus genes in the intestines of children with autism supported Wakefield's theories. (Eight years later, Mady Hornig and her colleagues at Columbia University's Mailman School of Public Health would discredit O'Leary's initial studies.)

Wakefield and O'Leary were followed by a cavalcade of doctors offering bogus cures. Mary Megson, a pediatrician and assistant professor of pediatrics at the Medical College of Virginia, said that she could cure autism with large doses of vitamin A. John Upledger, an osteopath from Palm Beach Gardens, Florida, explained that vaccines caused an abnormal flow of spinal fluid, which could be relieved by a manual decompression technique he had developed. Vijendra Singh, an immunologist from Utah State University, said that autism was caused by an autoimmune response against the brain. Michael Goldberg, a pediatrician from Los Angeles, claimed dramatic results with minerals, antivirals, antifungals, anti-inflammatories, and immune modifying agents. Burton was ecstatic. "That was a really good lecture," he enthused. "I enjoyed that, and we will have some questions about whether or not any of our health agencies have picked up on your procedures."

Not exactly a Capra-esque, Mr.-Smith-Goes-to-Washington moment. Didn't any of the committee members notice that these theories were mutually exclusive? Shouldn't someone have asked, "So which is it? Do vaccines cause gut abnormalities or vitamin A deficiency or an abnormal flow of spinal fluid or an autoimmune response against the brain? Is autism a viral infection or a fungal infection or a mineral deficiency? Which? Pick one. Because they can't all be correct." But no one questioned these obvious inconsistencies. Snowed by indecipherable jargon, they just kept nodding like a row of bobblehead dolls—as if all of this had made perfect sense to them.

I was on the next panel of scientists, the most important of whom was Brent Taylor, an epidemiologist from the Royal Free Hospital (ironically, the same hospital at which Wakefield had performed the research that had led to his later-to-be-retracted *Lancet* paper). Taylor had been the first to publish a study showing that children in England who had received the MMR vaccine were at no greater risk of autism than those who hadn't. Burton tried to challenge Taylor with information he had been given by his staff, but Burton was overmatched. Failing to put a dent in Taylor's conclusions, he eventually gave up.

I was up next.

Karen Muldoon Geus, the head of public relations at my hospital, was concerned about the controversial nature of the hearing, so she hired an outside public relations firm to work with me on my speech. The PR expert wasn't worried about what I would say; he was worried about my collaboration with Merck on the development of a rotavirus vaccine. Even though the vaccine was still six years from licensure, he wanted to make sure that my work with Merck was front and center in my speech. So I began my testimony with what probably sounded like a mea culpa:

> My name is Paul Offit. I am a practicing pediatrician. I am also the chief of infectious diseases and the Henle professor of immunologic and infectious diseases at the Children's Hospital of Philadelphia and the University of Pennsylvania School of Medicine, and a member of the Advisory Committee on Immunization Practices at the CDC. . . . My expertise is in the areas of virology and immunology. In addition, I have been in collaboration with Merck and Co. on the development of a rotavirus vaccine since 1992.

Despite this disclaimer, the PR guy was still worried. He wanted to know whether anything I had ever said or done could come back to haunt me. I told him that I had never been convicted of a crime, if that's what he meant. I said it jokingly, but he didn't laugh. He stared at me, as if looking through me, and asked again if there was *anything else out there*—anything that Dan Burton could use to discredit me. It's not hard to appeal to my (or I suspect anyone's) sense of guilt. So I thought back. Had I ever done anything illegal or immoral? I had smoked marijuana in college (and inhaled), but this was true of many people of my generation. I had never done harder drugs, like cocaine, heroin, or LSD. When I was eighteen years old, I had danced with Blaze Starr at a bar called the Red Rooster in Essex, Maryland. (Starr, Baltimore's most famous exotic dancer, was the subject of the movie *Blaze*, starring Lolita Davidovich and Paul Newman.) I couldn't imagine that this would be a problem, since it was *the single greatest moment of my adolescence.* During my junior and senior years in college, I had run a football gambling card that I had named "Pro Picks." Every week I would make my own point spreads on professional football games, print up the card, and distribute it to the dorms on campus. The advantage of my card was that it came out with point spreads a day before the Las Vegas bookmakers. The disadvantage was that I didn't know as much as the people in Las Vegas. (Although I certainly did try, spending a lot of time in the library looking through newspapers from NFL cities to find out who was injured.) Although this was no doubt illegal, it was harmless. I didn't tell the PR guy about any of these things because I couldn't imagine that he or anyone else would care.

Then I wondered whether events in my family's past could be embarrassing. One of my great-uncles had supposedly killed a man in his bar, but hadn't gone to jail. So I didn't think that would be a problem. There was, however, one thing that worried me. My uncle, Sidney Offit, a well-known author in New York City and curator of the prestigious George Polk Awards for journalistic achievement, had published a tell-all book five years earlier. *Memoir of the Bookie's Son* told the story of his father, Barney, whom everyone called Buck or Buckley. Buck Offit was without question the most successful of my grandfather's brothers and sister, even though he had never gotten past the fourth grade. From the time he had returned from service in World War I until the early 1950s, when the Kefauver Commission's investigation into organized crime put him out of business, Buck Offit was "among the elite of the nation's bookmakers." As depicted in movies like Sergio Leone's *Once Upon a Time in America* or Barry Levinson's *Bugsy* and *Liberty Heights*, Buck Offit thrived during a time when Jewish gangsters like Dutch Schultz, Legs Diamond, Meyer Lansky, Arnold Rothstein, and Louis "Lepke" Buchalter controlled organized crime in America. Buck certainly fit the part, spitting words out the side of his mouth like a Damon Runyon character. As did all of my family members, I idolized Uncle Buck. In the 1950s, Buck was arrested during a gambling raid. Although probably guilty of bankrolling the operation, he was able to bribe his way out of a conviction. I actually told the PR guy this story, even though I couldn't imagine Dan Burton or his staff digging through notes from the Kefauver Commission or reading Uncle Sidney's book. Not surprisingly, they hadn't.

In addition to Brent Taylor, the other scientists on my panel included Coleen Boyle, an epidemiologist at the CDC, who talked about autism rates; Edwin Cook, a child psychiatrist from the University of Chicago, who talked about behavioral therapies and the need for support services; and Deborah Hirtz from the National Institutes of Health, who talked about future research needs. Although I wasn't an autism expert, I had spent two decades studying viral pathogenesis, learning how viruses caused disease and how the immune system fought back. I thought I was in the best position to discuss the biological plausibility of Andrew Wakefield's claims about the MMR vaccine. Here's what I said:

> My role in these proceedings is to explore the theories that have
> arisen due to concerns by the public that autism might be caused

by the combination measles, mumps, and rubella vaccine known as MMR. No evidence exists that proves this association. However, three theories have been used to explain it. In the time that I have been given, I would like to explain why I think that these theories are invalid.

Everyone sat quietly. Staffers weren't leaning forward talking to their congressional representatives. Parents in the audience seemed to be paying attention. Members of the media stopped whispering among themselves. Great, I thought. Now finally I have a chance to explain to the public why this MMR–autism controversy hadn't made much sense:

The first theory is that children who get the measles vaccine make an immune response not only to the vaccine, but to their own nervous system. This kind of reaction is called autoimmunity. To understand why this theory is incorrect, we must first understand differences between natural measles infection and measles vaccination.

During natural measles infection, the measles virus reproduces itself many times in the body and causes disease. In contrast, following measles vaccination, the vaccine virus reproduces itself much less and doesn't cause disease. Because more measles proteins are made during natural infection than after immunization, the immune response to natural infection is *greater* than the immune response to immunization.

If the immune response is greater after natural infection, *then the autoimmune response would also be greater.* If this were the case, then autoimmunity should occur *more* frequently after natural infection than after vaccination. Or, said another way, if measles virus caused autism, measles vaccination would *lower*, not raise, the incidence of autism.

I thought this was an important point. I wanted parents to understand the tenets of autoimmunity without resorting to jargon, so I had worked hard on these sentences in preparation for my testimony. Then I addressed the next theory:

> The second theory is that the child's immune system is simply overwhelmed by seeing three viruses in a vaccine at the same time. Some have gone so far as to suggest that it may be of benefit to divide the MMR vaccine into three separate vaccines. [This was a direct shot at Andrew Wakefield.] The rationale behind this theory is that children do not normally encounter such an assault on the immune system. However, this notion is incorrect.

Wakefield glared at me from his perch about twenty feet away.

> From the birth canal and beyond, infants are confronted by a host of different challenges to the immune system. Their intestines encounter foreign proteins in milk and formula. Their lungs encounter bacteria inhaled on the surface of dust in the air. And literally thousands of different bacteria immediately start to live on the skin, as well as on the lining of the nose, throat, and intestines. So how does the infant deal with this immediate confrontation to the immune system?
>
> Babies have a tremendous capacity to respond to their environment from the minute that they are born. The newborn has billions of immunological cells that are capable of responding to millions of different microorganisms. By quickly making an immune response to bacteria that live on the surface of their intestines, babies keep those bacteria from invading their bloodstream and causing serious disease. Therefore, the combination of the three vaccines contained in MMR, or even the ten vaccines given in the first two years of life, is literally a raindrop in the ocean of what infants successfully encounter in their environment every day.

OK. Now I had only one more theory left. And it was probably the most difficult one to explain:

The third theory is that the MMR vaccine is given by an unusual route. The rationale behind this theory is that children normally inhale measles, mumps, or rubella viruses carried on droplets from another person, and do not normally have viruses injected under the skin. However, encountering viruses or bacteria under the skin or within the muscles *does* occur naturally. To meet this challenge, children have collections of immune cells in lymph glands located strategically throughout the body. For example, lymph glands located behind the elbow or under the arm. Because our skin can be cut, our bodies are ready to encounter challenges at any site.

Indeed, although wondrous, the birth process is quite traumatic. Newborns commonly have small cuts on their face and body after passing through the birth canal. Because the birth canal is covered with bacteria, the child will encounter bacteria under the skin immediately. Our species survives because, from the minute we are born, we are quite capable of meeting challenges at all sites.

Then I correctly predicted the future. Twice. Although anyone could have made these predictions:

> Parents testifying here today are asking a scientific question: "Does the MMR vaccine cause autism?" Questions of science are best answered by scientific studies. And the answer to this question is already available. Brent Taylor and his coworkers in London have conducted a large, meticulously designed, well-controlled study that disproved an association between MMR vaccine and autism. *I believe other studies will confirm Dr. Taylor's results.*

Sixteen confirmatory studies followed.

> My concern, and it should be the concern of this committee, Mr. Chairman, is that some parents listening to or reading about this hearing might incorrectly conclude that vaccines

cause autism. . . . If, as a result of reading about this hearing, some parents choose to withhold or delay vaccines for their children, their tragedy could be profound. *If many parents choose to withhold vaccines, the tragedy all across America could be devastating.*

In April 2000, when this hearing took place, measles had officially been eliminated from the United States. Fifteen years later, because enough parents had chosen not to give their children the MMR vaccine—in large part because they feared that it might cause autism—measles outbreaks swept across the country. Many of the children affected in these outbreaks were hospitalized.

When I finished speaking, I looked up at the panel expectantly. I thought I had done a decent job. So I was anxious to hear what the members of Congress thought. Burton was the first to speak. "Do you do any traveling around, speaking on behalf of Merck or Merck products?" he asked.

Wait? What? This is your question? Weren't you listening to what I just said? It wasn't until Burton asked this question that I understood what this congressional hearing was really about. This wasn't a scientific symposium, during which researchers share ideas and reach a consensus based on data. This was a courtroom, in which each side tries to win over the jury. In this case, the jury was the American public.

"I travel and speak about vaccines," I responded, "and these talks are supported by unrestricted educational grants from either pharmaceutical companies or from universities." Wrong answer. The right answer would have been, "No, I don't speak on behalf of Merck or any other pharmaceutical company." I had naïvely played the role of the teacher, trying to educate about how symposia are funded, instead of the role that I had been assigned at this hearing: witness. And, from Dan Burton's standpoint: hostile witness. I had also failed to realize that Burton probably had no idea what an unrestricted educational grant was, no idea that the word *unrestricted* meant that the company had no say in who spoke or what was said. My realization of these things came far too late to save me. I was just treading water, trying to survive Burton's staccato prosecutorial style.

"So they pay for your expenses and that sort of thing?" said Burton.

"They have an interest in educating physicians about vaccines," I said, "and it is good that they do, because physicians need to be educated about vaccines." The truth is, I didn't really know who supported these medical and scientific symposia. I had always assumed that it was hospitals or universities or government grants or unrestricted educational grants from pharmaceutical companies, but I wasn't sure. In any case, I knew that whoever funded them had no influence on what I said. It wasn't until he asked this question that I realized that Burton had invited me to the hearing to paint me as a shill for Merck, despite my opening disclaimer. Burton said, "I understand. And they [Merck] produce the MMR vaccine, don't they?"

"Yes," I said. "They do, yes."

"Thank you," said Burton.

I felt terrible. I thought I had made it clear that I had been collaborating with Merck on the development of a rotavirus vaccine. I thought I had been transparent. But now Burton had implied that there was something more sinister going on. That I was hopping around the country promoting Merck products, which wasn't true. I had worked hard on my speech. But after Burton's questions, I felt dismissed, dirty. It was awful.

In retrospect, I realize I had been the victim of the old legal aphorism that when the law is on your side, argue the law; when the facts are on your side, argue the facts; and when neither are on your side, attack the witness. Burton couldn't question my conclusions because he didn't know enough to question them— so he attacked me personally.

Then one of the Democrats on the committee tried to help.

John Tierney, a Democrat from Massachusetts, upset at the way the hearing was going, gave Coleen Boyle, the CDC epidemiologist, and me a chance to stand up for vaccines, asking, "If you had young children today, would you vaccinate them?"

"I do have children," Boyle said. "And they are both fully vaccinated. And I would vaccinate them again."

"Dr. Offit, how about you?" asked Tierney.

"Yes," I said. "I have a seven-year-old son, Will, and a five-year-old daughter, Emily. And they are both fully vaccinated."

"And you would do it again?"

"Of course," I said. "I want them to be protected against the viruses and bacteria that can cause serious disability and death. I am fortunate, actually. I was a little boy in the 1950s. And when I was a little boy, there were four vaccines: diphtheria, pertussis, tetanus, and smallpox. I was fortunate that I wasn't killed by measles or paralyzed by polio. My son did not have to be as lucky."

It didn't take long for me to find out that once again I had screwed up. At the break, one of the Democratic staffers came up to me and grabbed my arm. Pulling me close to him, he said, "Never, never mention the names of your children at a hearing like this!" Having no experience testifying in front of congressional committees, I didn't know where the potholes lay. So I kept falling into them. I had never considered that, by mentioning my children's names, I might have been putting them at risk.

Later that morning, Burton claimed that my collaboration with Merck should have disqualified me from something far more important than speaking to his committee. The revelation occurred during his questioning of Coleen Boyle.

"I want to ask Dr. Boyle one last question," said Burton. "And that is, Do you believe anybody who is getting funds from Merck or any of the pharmaceutical companies should be on advisory panels that are making judgments about pharmaceuticals coming from those companies? Or do you believe that it is a conflict of interest?"

Boyle hesitated. "I think that is a difficult question to answer," she said.

Burton pressed his advantage. "Wait a minute!" he shouted. "Let me get this straight. You think it is a difficult question to answer? If somebody is getting funding of some type from a pharmaceutical company, for them to sit on an advisory panel that is approving or giving their approval to a new drug that is coming on the market, you do not see that as a conflict?"

Ben Schwartz, who was the associate director of the CDC's Epidemiology and Surveillance Division and had an intimate knowledge of how the Advisory Committee on Immunization Practices (ACIP) worked, came to Coleen's rescue. (Voting members of the ACIP decide which vaccines should be recommended for U.S. children.) "There are very strict guidelines regarding the participation in votes of members who may have conflicts of interest that will help assure that those potential financial conflicts do not affect the votes and the decisions of the advisory committee," said Schwartz. "The reason why individuals who may potentially have conflicts are included in the committee is to assure that we get the best expert advice possible so that we can make the best vaccine recommendations possible."

Dan Burton and Ben Schwartz were talking about me. I had joined the ACIP as a voting member in 1998 and had remained on the committee until 2003. The Burton hearing took place in 2000, so I was still on the committee at the time. I was brought on to the ACIP because of my expertise on rotaviruses. At that point, I had already published about eighty papers on the subject and was one of only a handful of experts in the area of rotavirus-specific immunity. Because the ACIP was on the verge of recommending a rotavirus vaccine made by Wyeth, the CDC thought it would be of value to include me on the committee. To Dan Burton, this was a clear conflict of interest. But what was the conflict? I had voted *for* recommending Wyeth's vaccine, not against it. Also, and this was my fault, I should have made it clear to the Burton committee that I had *never received financial compensation from Merck*—not in support of my laboratory, not in support of talks, not in support of advice, and not in support of my salary. My funding had come solely from the National Institutes of Health, the Children's Hospital of Philadelphia, and the Perelman School of Medicine at the University of Pennsylvania. Burton assumed that when I said that I had worked "in collaboration with Merck," it meant that I had been financially compensated to do so. I should have clarified the difference between collaboration and compensation. But I suspect that it wouldn't have mattered. Burton would probably have believed that I was lying about my funding no matter what I had said.

Dan Burton never gave me a chance to respond to his claim that I was in the pocket of industry and, as a consequence, had served my own interests at the expense of the interests of children. But Henry Waxman did. A few minutes after Burton grilled Coleen Boyle, Waxman had the floor. "Now, Dr. Offit's integrity has been challenged," he said, "presumably because he has a point of view that does not quite fit with the chairman's point of view. Dr. Schwartz started to indicate why he thought your situation, even though you have a relationship with Merck, did not put you in conflict. Let me ask you directly, Dr. Offit, do you have a conflict of interest, and, if no, why not?"

There's an old saying that a Republican is a Democrat who has been mugged, and a Democrat is a Republican who has been indicted. I had felt like I had been mugged *and* indicted. But at that moment, I loved the Democrats.

"I have no conflict of interest," I said. "What I have is an *apparent* conflict of interest, and that is why I disclosed that at the beginning of every ACIP meeting, and that is why I disclosed it in my written report [to the Burton committee]. If I could just explain this a little bit. I have been doing research for twenty years on rotaviruses. What I have done in my laboratory is try, with my colleagues, to understand what [rotaviral] genes cause diarrhea and what [rotaviral] genes help the body fight infection. . . . Rotaviruses cause one of every seventy-five children born in this country to be hospitalized. It is a serious and, in developing countries, often a deadly infection. It would be an advance if we could prevent that disease. Merck and Company has made a commitment to developing that vaccine and, hopefully, if we can develop a safe and effective vaccine, we can prevent a lot of disease and death."

This was a terrible answer. Most important, I didn't answer the question. What I had tried to do (idiotically) was to defend the manner by which vaccines come into existence, to explain that only pharmaceutical companies have the resources and expertise to make vaccines and that we shouldn't vilify them for doing it. Waxman had asked me to defend myself and instead I had defended the process.

Then Waxman gave me a second chance. "I think that everyone here should agree that we want a safe vaccine," he said, "or a vaccine that is as safe as possible. Merck did not hire you to come up with a particular position, did they? Did they tell you they wanted your research to have a certain outcome?"

I said, "This work was all . . . funded by the National Institutes of Health, which funded my basic research. I am an immunologist. That is my expertise." Better, but still not very good. The better answer would have been, "I have never received money from Merck. And I wish the chairman would stop saying that I have. Frankly, if the chairman had any idea why the NIH- and CDC-funded researchers in this room do what they do, he would stand up right now and apologize." (Don't you wish life could work this way? That you could go back in time and give the kind of impassioned speech that only actors in scripts written by Aaron Sorkin ever manage to pull off?)

It seemed to me that all the work I had done on my speech had been a waste of time. After the hearing ended, when I was standing at the back of the committee room, one mother came up to me to confirm my disappointment, saying that she hadn't realized that I was "such a jerk." But then a few other parents came up and asked me to explain in more detail why I didn't think that the MMR vaccine could cause autism. They had listened carefully to what I had said and asked good questions. I felt sorry for these parents. They had come to Dan Burton's committee hearing to get answers, hoping to learn about the latest research on the cause or causes of autism and about what could be done to help their children. Instead, they had been subjected to seven hours of testimony that did nothing but serve the financial interests of hucksters offering a bazaar of alternative therapies and the biases of an ill-informed and frankly vindictive committee chair.

CHAPTER 10

The Nuclear Option

What is objectionable, what is dangerous about extremists,
is not that they are extreme, but that they are intolerant.
The evil is not what they say about their cause,
but what they say about their opponents.

—ROBERT F. KENNEDY (FATHER OF RFK JR.)

Anti-vaccine activists like Jim Carrey, J. B. Handley, Robert F. Kennedy Jr., Barbara Loe Fisher, Jenny McCarthy, Lyn Redwood, and Andrew Wakefield often say that they're not anti-vaccine; they're pro-vaccine. They just want safer vaccines. Like them, I, too, want safe vaccines.

Indeed, when I first encountered a real vaccine-safety activist, I did everything I could to help him. His name was John Salamone. John's son, David, had developed polio from the oral polio vaccine. Although this side effect is rare—occurring in about 1 in 2.4 million doses—it's real. In response, John founded the organization Informed Parents Against Vaccine-Associated Paralytic Poliomyelitis. The difference between John Salamone and anti-vaccine activists is that Salamone had identified a legitimate vaccine safety issue. Anti-vaccine groups, on the other hand, want vaccines that don't cause allergies, asthma, autism, chronic fatigue, diabetes, multiple sclerosis, rheumatoid arthritis, or sudden infant death syndrome, among others. Because vaccines don't cause these disorders, they're asking for something they already have.

In 1998, when I first joined the Advisory Committee on Immunization Practices (ACIP) at the CDC, I was made head of the polio vaccine working group. At the time, children in the United States were receiving the oral polio vaccine, the one that had partially paralyzed John Salamone's son. I had seen John at national meetings of pediatricians and, quite frankly, he hadn't been treated particularly well. He was often asked to place his organization's booth, which included several children who had been harmed by the vaccine, in the most inaccessible spot in the exhibition hall (like one of those remote circus cages in Franz Kafka's *A Hunger Artist*). I asked John if he would be willing to join the CDC working group as a parent representative. Within two years, we eliminated the oral polio vaccine from the United States; now, because we use only the inactivated vaccine, which is given as a shot, the risk of getting polio from the vaccine has disappeared. No longer do children have to suffer as David Salamone suffers. By putting a human face on the problem, John Salamone had made vaccines safer for children. Because I had led this effort, I, too, consider myself to be a credible vaccine safety activist.

Nonetheless, anti-vaccine activists don't see me that way. And they do everything they can to discourage me from speaking out.

• • • •

IN THE MOVIE *THE GODFATHER PART II*, THE CHARACTER HYMAN Roth tells Michael Corleone that he didn't complain after the Corleone family had killed his friend Moe Green. "This is the business we've chosen," Roth says. Similarly, if scientists are going to try to educate the public about an aspect of science about which people are passionate, we must accept that things can get ugly. And the way that it gets ugly is with threatening emails, physical harassment, frivolous lawsuits, and death threats.

I'll start with the emails.

Whether the controversy is climate change, evolution, GMOs, alternative medicine, dietary supplements, or vaccines, the themes of hate email are always the same: money, murder, Nazis, God, Satan, and Judgment Day. For example: "You are a demon straight out of hell"; "You are a very sick and evil man"; "You, sir, have blood on your hands"; "Your day of reckoning will come"; "I just pray that the love of Christ will one day flood your darkened heart"; and "You should be ashamed of yourself and I hope you're ready for Hell, because I'm positive that when you finally croak, Satan will have his own special place in Hell for evil people like you." Sometimes the emails are more hopeful, offering a chance at redemption: "Do the right thing. Turn away from evil and corruption. You are a disgrace to humanity. You can change."

Occasionally, emails are sent to someone else but refer to me. For example, Perri Klass, a professor of journalism and pediatrics at New York University and the director of graduate studies at the Arthur L. Carter Journalism Institute, received an email with the subject line "Nazi Bitch Whore": "Shut the fuck up if you don't want to be put next to Offit in the firing squad once your tyranny collapses." (I suppose it shouldn't be surprising that an email whose subject line is "Nazi Bitch Whore" contains a message that isn't particularly supportive.)

In truth, these emails always upset me. I have yet to become thick-skinned enough that they don't hurt. So I've developed strategies to distance myself from them. One is humor (or at least what passes for humor), which helps a little. Another is denial. I simply refuse to believe that people really hate me that much. I think that if they actually knew me, they wouldn't feel that way.

Although I have been tempted on occasion to answer these emails, I know it would be a mistake. After my first response, I would be far more intimately involved with someone I definitely don't want to be intimately involved with. So I try my best to ignore them. Reasonable people aren't writing them. So logic and common sense aren't going to change their minds.

• • • •

I'VE ALSO BEEN SUED.

My guess is that the purpose of the lawsuits isn't to win; it's to get me to spend my time and money—and mostly to shut me up. If they can make speaking out on behalf of science onerous enough, then presumably I'll stop.

On December 23, 2009, Barbara Loe Fisher, the director of an anti-vaccine group called the National Vaccine Information Center, sued me for libel. Her claim was based on an article written in the magazine *Wired* titled "An Epidemic of Fear: One Man's Battle Against the Anti-vaccine Movement." Amy Wallace, a celebrity journalist, wrote it. Wallace is probably best known for an article she wrote in *GQ* about Charlie Sheen titled "Charlie Sheen's Demons: Coke, Hookers, Hospital, Repeat." The teaser for that article read "Charlie Sheen talks to Amy Wallace about his latest bender, his true feelings about sobriety and 'Apocalypse Now,' and the cyclical insanity of his crazy-ass life." I can only imagine how much of a comedown it must have been for Amy to have had to interview me.

In her article, Wallace talked about the ill-founded fears of anti-vaccine activists, the cottage industry of false cures for autism, and the celebrities who support the movement. The article was sympathetic to me and to the science. The paragraph that got me, Amy Wallace, and Condé Nast (the publisher of *Wired*) sued for $1 million was "Fisher, who has long been the media's go-to interview for what some in the autism arena call 'parents' rights,' makes [Offit] particularly nuts, as in 'You just want to scream.' The reason? 'She lies,' he says flatly. 'Barbara Loe Fisher inflames people against me. And wrongly. I'm in this for the same reason she is. I care about kids.'"

The sentence to which Fisher objected most was " 'She lies,' he says flatly." I found out about the lawsuit when Barbara Loe Fisher's lawyers sent a process server to my home. People can learn that they've been sued either by certified mail or by having a guy (who in my case looked like a bouncer at a rough nightclub) come to your house, knock forcefully, and then loudly proclaim that you are being served. This is the more expensive and more embarrassing way, especially if you have guests, and it's the way Fisher chose. My wife answered the door. After the man handed her the papers, she chirped brightly, "Honey, we're being served." (My wife has far too much fun with this. She, like I, feels that this isn't really our life. The truth is that we're not all that interesting and, therefore, aren't the kind of people who should be getting served with anything. I agree with her. I feel that the anti-vaccine people have created an anti-hero, a villain, and they want me to play the role of the villain. I counter this by imagining that I'm in a musical farce about dancing milkmaids from Ukraine.)

When I learned that I was being sued, I called the lawyer who was representing Amy Wallace to see if Condé Nast's insurance company would cover my legal fees. The Condé Nast lawyer explained that publishers don't typically cover people who are interviewed for a story; they cover only the writer. He also said that if the lawsuit got to the point that each side would be required to give a deposition, it could cost me between $100,000 and $200,000; if it went to trial, it could cost me as much as $1 million (which, ironically, was the amount for which I was being sued). He reassured me that Fisher's lawsuit was baseless and that in a just world, it would be dismissed. At issue was whether we were living in a just world.

Then I got some good news. Miraculously, the umbrella policy on our homeowners' insurance covered me for libel. My insurance company turned the case over to a law firm in the county in Virginia where Fisher had brought her lawsuit. On March 10, 2010, the case was dismissed. Regarding me, the judge ruled that "a remark by one of the key participants in a heated public debate stating that his adversary lies is not an actionable defamation," but rather "an expression of opinion" protected by the First Amendment. Regarding Condé Nast and

Amy Wallace, the judge ruled that "the impassioned response by Defendant Offit toward Plaintiff was itself illustrative of the rough-and-tumble nature of the controversy over childhood inoculations and therefore worthy of mention in the *Wired* article."

Even if the judge had decided that my quote was not an opinion, but rather a statement of fact, Barbara Loe Fisher would have had to prove that my statement "She lies" was incorrect. Second, she would have had to prove that I had acted with actual malice, meaning that not only was I wrong about her, but that I knew I was wrong, or had at least shown a reckless disregard for the truth. This, too, was a high bar; it's hard to prove intent. Finally, since Fisher was suing me for $1 million, she would have had to prove that my statement had substantially damaged her reputation. I wasn't the first person to say that Barbara Loe Fisher was a source of potentially harmful misinformation. So again, proving that I had damaged her reputation beyond the degree to which it had already been damaged would have been hard.

One thing I learned during the Fisher lawsuit was that it's good to live in a country that has constitutionally protected free speech. This wasn't the case when a British lawyer named Richard Barr threatened me with a lawsuit. In England, the burden falls on the defendant to prove that what was said was true, not on the plaintiff to prove that it wasn't. Barr was the lawyer who had funneled money from England's Legal Services Commission to Andrew Wakefield when Wakefield claimed that the MMR vaccine caused autism. Barr's threatened lawsuit was based on a statement in my book *Deadly Choices: How the Anti-Vaccine Movement Threatens Us All*. Referring to the investigative journalist Brian Deer, who had blown the cover off Wakefield's misrepresentations, as well as his source of funding, I had written, "Deer also found that the personal injury lawyer who represented these children, Richard Barr, had given Wakefield £440,000 (about $800,000) to perform his study, essentially laundering legal claims through a medical journal." Barr claimed that I had made it sound like he had paid Wakefield out of his own pocket, when in fact he had just directed the money from the Legal Services Commission. According to the lawyers representing Basic Books, which had published *Deadly Choices*, this kind of lawsuit would never succeed in the United States, the argument being that no reasonable reader would assume that Barr had pulled $800,000 out of his own pocket. But the lawyer also warned that in England, the burden of proof isn't on the plaintiff to prove wrongdoing; it's on the defendant to prove no wrongdoing—in essence, to prove that no reader could possibly interpret the sentence the way Barr had claimed. We eventually removed that sentence from the British edition of the book.

In the midst of this back-and-forth with the publisher's lawyer, I called Richard Barr. As crazy as it might sound, I really liked him. He seemed burned by his association with Andrew Wakefield and was trying to reclaim his reputation following Wakefield's dramatic fall from grace. (Wakefield's paper had already been retracted, and he had lost his license to practice medicine.) We agreed that if I sent £1,000 (then about $2,000) to a charity that provided services for children with autism, he would withdraw his lawsuit. A happy ending.

Probably the best advice on how to handle these lawsuits comes from Ken White, whose online nom de guerre is Popehat. On September 26, 2013, White wrote an article titled "So You've Been Threatened with a Defamation Suit." "Hi, I'm Ken White," he began. "You may remember me from such defamation-related posts as 'You Can't Call a Bigfoot Hunter Crazy, That's Libel!' and 'If All Critics of Dentists Go to Jail, Then Only Criminals Will Criticize Dentists!' " White offered advice for what to do if you're being sued for something that you have written (libel) or said (slander). The first piece of advice was to calm down. White called this the "Martha Stewart Rule." "Lots of people get in trouble not because they did something wrong, but because they heard they were being *investigated* for doing something wrong, and they panicked and started lying and deleting files and setting cabinetry on fire and making angry statements to the press."

Next, White advised his readers to gather important information. Who threatened to sue you? Was the person who contacted you the one who was suing you or a lawyer working on their behalf? Has the lawsuit already been filed, or is this just a threat? White also strongly encouraged people to educate themselves about what you can and can't say about other people. "One of the most important concepts in defamation law is that *statements of fact* can be defamatory ('You got drunk and ran over my polecat with your Segway!')." In other words, if you say or write something that is true but subjects the other person to hatred, ridicule, or contempt, then you could be found guilty of defamation.

White pointed out that for the most part, in a country with a First Amendment, you have free rein to state your opinion about a public figure like Barbara Loe Fisher: "Pure opinion ('You're the worst dad') and 'rhetorical hyperbole' ('You're an asshole traitor)" are not defamatory. In other words, I could have called Fisher things far worse than a liar.

· · · ·

ANOTHER STRATEGY USED BY ANTI-SCIENCE ACTIVISTS IS harassment.

I've been stalked, grabbed, shouted at, and received some pretty harrowing phone calls. The goal of these attacks is to get me to react—to say or do something that will discredit me. I'll give you an example.

On November 21, 2016, I was asked to participate in a symposium at the NYU Langone Medical Center in New York City. The title of the symposium was "Confronting Vaccine Resistance: Strategies for Success." Richard Pan, the state senator who had led the fight to eliminate California's philosophical exemption to vaccination; Dorit Rubinstein Reiss, a law professor and vaccine advocate from the University of California Hastings College of the Law; and Benard Dreyer, the president of the American Academy of Pediatrics, had also been asked to speak.

A few days before the meeting, activists issued a call to arms on a prominent anti-vaccine website:

> Please plan on attending a demonstration against an event entitled "Confronting Vaccine Resistance." . . . The meeting is closed to the public but will feature three of America's leading proponents of forced vaccination. . . . Bring friends, families, and cameras. And bring a poster with a picture of any vaccine-injured loved one, along with their name, the date they were injured, and the vaccine(s) that injured them.

A few years earlier, anti-science activists had staged a similar event at the CDC in Atlanta. In order to get to the meeting, I had to walk through a group of angry protesters carrying signs, one of which showed a picture of me with a bright red slash across my face above the word "TERRORIST!" While walking through the crowd, one of the protesters grabbed my arm and spun me around. I asked him to please let go, and eventually he did. But it was frightening. I had no interest in repeating that experience. So I got on an early train from Philadelphia, arriving in New York City three hours before the event started. When I entered the main building, I met with the chief of security, who was well aware of the protest. I asked him if I could wait in the meeting room. He said that the room wasn't ready yet, but that I could wait in the cafeteria. "You'll be safe there," he said. "Relax, have some breakfast. No one will bother you. When it's time for the symposium to start, we'll escort you to the meeting room."

The cafeteria was located on the first floor of the hospital. I bought a bagel and juice and took a seat by the window (not a wise choice). After a few minutes, I noticed a woman with a large camera filming me from the street. I recognized her as Polly Tommey—a collaborator of Andrew Wakefield—who believes that all vaccines are unsafe and advises parents to be wary of pediatricians. Standing next to her was another woman pointing toward a large black bus with the word "VAXXED" printed on the side, underneath of which was the phrase, "Where There's Risk, There Should Be Choice." *Vaxxed* was an anti-vaccine movie that had been written and directed by Andrew Wakefield.

I couldn't believe that someone was filming me from the street. At this point, a security guard walked up to the window and frantically waved at Polly Tommey to stop. I was about to get up and move away from the window when a man who appeared to be in his early forties with a camera hanging around his neck came up behind me. The camera had a large lens that came close to my face. He asked if I would like to come onto the bus to be interviewed by the *Vaxxed* team. If I had said yes, I would have had to go outside and talk to a group of people who not only didn't care what I had to say but routinely vilified me on their websites. If I had said no, it would have sounded like I didn't care about parents who only wanted a chance to express their concerns.

I said no. Then I asked him who he was. He said, "I'm the cameraman for *Vaxxed*." He asked again if I would come talk to the *Vaxxed* parents. I said, "You've got to be kidding." At this point, I should have gotten out of my chair and walked away. But I was scared, and also a little angry. I felt ambushed. Plus, the camera in my face made it difficult for me to stand up. (I assumed that his camera took photographs. I hadn't imagined that the camera was filming me while he stood there with his hands at his side. Yes, I'm that old.) The young security guard waving to Polly Tommey wasn't helping. The cameraman said that if I had truth on my side that I should want to come out and talk to the parents. What was I afraid of?

The cameraman wasn't going away and, although he appeared calm on the surface, he was seething. I was trapped in my chair. So I said, "Get out." Then I said the magic words that would haunt me (and my hospital) for the next few days. I said, "Get the fuck out." I regretted it the minute I said it. Undeterred, he again asked me to come outside. I said, "Out!"

Within minutes, a video of me cursing at this man appeared on an anti-vaccine website and soon on YouTube (where, by the end of the day, thousands of people had seen it). Later, I was escorted to the meeting by a security guard. A series of security guards never left my side for the rest of the day. When the symposium started, it seemed that everyone in the room had seen the video. The head of public relations at NYU Langone said that she would be happy to provide me with media training (assuming that I didn't know that you weren't supposed to curse at people who were filming you).

At the time that I was harassed, I didn't know the man who had confronted me. Later that day, Dorit Reiss informed me that it was Joshua Coleman and directed me to a news article dated October 30, 2015, titled "Roseville Anti-vaccination Campaigner Charged with Willful Cruelty to Children." Roseville is in California. Joshua Coleman had apparently traveled from California to New York to attend this protest. Coleman had been arrested for keying a car parked in a handicapped parking space, even though the car had a handicapped sticker. While he was vandalizing the car, he was holding one of his young children under his arm. When the police were informed of the vandalism and tried to arrest Coleman, he ran away and hid in a garbage can. "I've found people in trash cans before," said Sergeant Jason Bosworth of the Roseville Police Department. The article stated that Coleman was "now facing charges of vandalism, willful cruelty to a child, and obstructing a public official." This wasn't Coleman's first angry outburst. Coleman had also been shown on a Facebook post harassing Dr. Richard Pan, the California state senator who had spoken at the symposium in New York City.

This article helped me understand why Coleman had scared me. Coleman hadn't been loud and confrontational, but he was clenching his teeth when he spoke, clearly holding back his anger. He wasn't Charles-Manson-blue-swastika-on-the-forehead scary. He was more of a placid-disarming-smile-of-someone-who-might-hurt-you scary. So I struck back more out of fear than anger. But what I did was unacceptable. I had given the anti-vaccine activists exactly what they had wanted. I had made their day.

When I posted the details of the episode on my Facebook page, people who have supported my efforts responded. Most wrote letters to my hospital's director of public relations, concerned that the hospital wouldn't support me. Frankly, the responses were more than I felt I had deserved. For example:

From a microbiologist at the CDC:

In October, the anti-vaccination and anti–public health group calling itself "Vaxxed" protested outside the main gates of the CDC. I was working that day and went to lunch, as I normally do on Friday, with a group of fellow scientists. As we left the security checkpoint, we were surrounded by women and children shoving signs in our faces, yelling at us from multiple angles and accusing us of all sorts of heinous acts. They shoved cameras in our faces without permission, stood in our way as we moved forward, and tried to incite us to say something to the salacious things they threw at us. Thankfully, we were able to get past them as they screamed on a bullhorn, "We can smell your souls rotting from here; you are going to HELL." This group is simply trying to bully and push people and catch them in a moment of frustration when they do NOT leave after repeatedly asking to be left alone. They don't film that part. I wanted to bring this to your attention as I think their tactics when they attacked Dr. Offit were the same. Please consider the source.

From a professor at the University of Edinburgh in Scotland: "I am writing to assure you that you are exposed to what is a coordinated activity of an exceedingly small, albeit very vocal, minority of vaccine refusers. The vast majority of parents and patients are tremendously grateful for Dr. Offit's work as a doctor, inventor, and public health/vaccine advocate. His, and by proxy, CHOP's, excellent reputation extends across and beyond the U.S."

And from a mother of a young toddler (and my personal favorite): "I, along with many others, was so glad to see Dr. Offit tell 'Vaxxed' to get fucked."

There is an old saying that you should never find out how much your friends are willing to stand behind you. And while it was certainly heartening that my hospital received dozens of emails and letters supporting me, I had tested my friends unnecessarily. And you can only push that button so many times.

One of the things I never understood was why these protesters did what they did. There they were, across the street from the NYU Langone Medical Center with their signs, standing out in the January cold for hours. Several had infants and young children with them. It reminded me of the CDC protest at which a mother of a five-year-old boy had dressed him in a T-shirt that read "Damaged by Mercury." What kind of a parent has their child wear a shirt with the word "Damaged" on it? Also—and I only wish I had taken a picture—the mother was smoking a cigarette at the time.

These parents represent a very small minority of parents of children with autism. Although they often get most of the media attention, it's the other parents of children with autism who are the real heroes. Parents like Alison Singer, a former NBC executive producer and the mother of a daughter severely affected by autism. Alison founded the Autism Science Foundation (the only autism foundation with the word "Science" in the title), which is devoted to raising money to fund autism research. It's unlikely that this research will help her daughter. In that way, Alison is remarkably selfless, giving so much of her time and energy to help future children. Or parents like Matt Carey, Shannon Des Roches Rosa, Liz Ditz, Fiona O'Leary, and Ken Reibel, who fight for services for these children, as well as to get people to appreciate what children with autism have to offer and to destigmatize the disorder.

But what exactly did the *Vaxxed* parents standing in front of NYU Langone hope to accomplish by harassing scientists and public health advocates? How did it help their children? In 2015,

I spoke at a symposium at the University of California, Berkeley. While speaking, a woman in the second row held up her iPhone to film me. I asked her to please stop, but she wouldn't. So I gave up. At the end of the talk, she walked to the front of the room and put her camera phone inches from my face. I asked her to please put the camera down, but she wouldn't. So I walked away. She wanted me to get angry, to touch her or her camera so she could claim battery—that I had touched her against her will. The same thing happened at NYU Langone. The parents innocently claimed that all they wanted was for me to come outside and talk with them. But they didn't want that at all. They only wanted to upset me—to make me suffer. My mistake was that I let them get to me. As one supporter wrote, "Dr. Offit is a scholar and a humanitarian and a gentleman. But he is not a saint." True enough.

. . . .

THEN THERE ARE THE DEATH THREATS.

Not too long ago, I received an email that read, "If someone is forced to vaccinate their child when they did not want to because of your legislated ideas, and that child dies, then that parent should have the right to kill you. So please, continue your push for universal vaccination." Although this email contains an implied death threat, it doesn't reach the level of an actionable threat. The following story explains why.

In the early 2000s, I received an email from a man in Seattle that read, "I will hang you by your neck until you are dead." Every day in the United States, thousands of threatening emails are sent. In order for an email to reach the level of an actionable death threat, three criteria must be met: (1) The threat must be specific. It can't read, "Hope nothing happens to you," or "Watch your back." By stating *how* he would kill me, the Seattle man had met the first criterion. (2) The threat must be made more than once. Sadly, I received another threat from the same man. (3) The threat must be made by someone who might actually do it; for example, a person with paranoid schizophrenia.

When my tormentor met all three criteria, an FBI agent in Philadelphia set the wheels in motion to investigate him. The FBI told the man to "cease and desist." Then agents checked his emails, monitored his phone calls, checked to see whether he had purchased weapons, and observed his comings and goings. All of this was very reassuring. (By the way, any high-minded notions you might have about civil liberties vanish the minute you're threatened. I had no problem with law enforcement officials invading this man's privacy. Please, invade away.)

When the Seattle man threatened me, I was still a voting member of the ACIP. This meant that six days every year, I worked for the federal government. Because the threat was related to my public statements about vaccines, and because I worked for the government in the area of vaccines, the CDC offered me protection for a few ACIP meetings. At the time, these meetings weren't held on the CDC campus—where guards and metal detectors are at every entrance—they were held at a Marriott hotel, where there was little security. So a sheriff was assigned to watch out for me. He would follow me to and from lunch, a gun at his side.

. . . .

SADLY, ALL OF THESE TACTICS OF INTIMIDATION HAVE FORCED A number of good scientists to the sidelines.

CHAPTER 11

Pharma Shill

*Never dare to sell your soul for money, because no amount
of wealth will buy you an air conditioner in hell.*

—EDMOND MBIAKA

One strategy used by anti-science activists to discredit
scientists is to claim that scientists are in it for the money—
that they are simply parroting the message of a for-profit
industry.

In 2013, while standing on the steps of the Capitol Building,
Robert F. Kennedy Jr. said that I "should be put in jail, and the
key should be thrown away." Then he called me a "biostitute."
(I'm not exactly sure what that means, but it didn't sound good.)
Jim Carrey, in an article written for the *Huffington Post*, called
me a "profiteer." Jenny McCarthy tweeted, "Lemon girl scout
cookies recalled, so says tweets. Can't we recall bad people?
Like Paul Offit?" To which Chelsea Handler responded, "Who
the fuck is Paul Offit?" (My children loved this. They couldn't
believe that Jim Carrey, Jenny McCarthy, and Chelsea Handler
had taken time out of their busy, celebrity-filled lives to vilify
me. It was one of those rare moments when your children look
at you with genuine pride.)

Indeed, not a week goes by when I don't receive at least one mean-spirited personal attack that includes the phrase "Pharma shill" or "Pharma whore." For example: "You are the spawn of Satan. I don't care about your cars, your money, or anything else. But I'll be damned if you think you can take away my rights so you can buy the world."

To explain what I have done, why I have done it, and why I persist in trying to educate the public about science, I'll start at the beginning.

I was born at Sinai Hospital in Baltimore, Maryland. My father manufactured men's shirts, and my mother taught adult education in the Baltimore City school system. Despite claims to the contrary, nothing about my birth suggests that I was the spawn of Satan. When my mother first held me, she didn't scream, "What have you done to it!? What have you done to its eyes!?" (Sorry. Everything I know about spawns of Satan comes from the movie *Rosemary's Baby*.) My father didn't turn away, knowing that he wasn't my real father, knowing that he had made a deal with the devil. And no one in the delivery room shouted, "Satan is his father! He came up from Hell and begat a child of mortal woman. He shall redeem the despised and wreak vengeance in the name of the burned and the tortured." At least not as far as my mother can remember. When I had my DNA checked by Family Tree DNA, I found that my background is mostly eastern European, as is that of my parents and their parents and their parents. Plus, I look like my father. All of which leads me to conclude that I'm not the spawn of Satan, Prince of the Underworld. I'm the spawn of Shirley and Morton Offit, a middle-class couple from suburban Baltimore. They also spawned my brother and sister.

• • • •

WHEN I WAS FIVE YEARS OLD, TWO EVENTS SHAPED MY LIFE.

While playing on a slide one day when I was in kindergarten, I fell about twelve feet, landing face down. I lay at the base of the slide for about thirty minutes, not moving. Eventually, one of the other children went inside to find a teacher. When I was more easily aroused, the teacher put me on a school bus and sent me home. That afternoon I complained of pain in the upper left side of my abdomen. So my mother took me to the pediatrician's office. Unfortunately, my regular pediatrician, Dr. Milton Markowitz, was out of town. One of his partners examined me, but couldn't find anything wrong.

One day passed.

The following evening, Dr. Markowitz returned. After sifting through his partner's notes, he called my mother to see how I was doing. She said that I was still complaining of pain. So he drove to our house, examined me, and explained that I had ruptured my spleen and required immediate surgery. My father called my grandfather, who came to the house and asked Dr. Markowitz if he wouldn't mind getting a second opinion. Dr. Markowitz said that there wasn't time and that he would drive me to the hospital himself. I sat in the back seat next to my grandfather. (I remember being excited that I got to ride in a car at night in my Baltimore Colts pajamas.) The surgeon found a ruptured spleen, a quart of blood in my abdomen, and impending shock. Dr. Markowitz had saved my life. Had he not come back to the office that night, reviewed his partner's notes, called my mother, driven to our house, insisted on surgery, and resisted my grandfather's request for a second opinion (which would not have happened that night), I would in all likelihood have died in my sleep.

I idolized Dr. Markowitz; I wanted to be just like him. When he came to our house, he always took time to show me all of the marvelous tubes, devices, syringes, and medicines in his big black bag. After tapping on my chest, he tapped on the wall above my bed to demonstrate how the sound changed from hollow to dull, indicating the presence of a stud. This, he explained, was the principle he used to see if I had pneumonia. My mother often took my sister, brother, and me to Dr. Markowitz's house to get our vaccines, which he pulled out of his refrigerator. Vaccines were a family affair: our family and Dr. Markowitz's family.

I will always remember Milton Markowitz as the kind, generous, thoughtful hero who had saved my life. This was my image of a pediatrician, which no doubt played a role in my choosing pediatrics as a specialty. Dr. Markowitz eventually left private practice to pursue a career in academic medicine, rising through the ranks to become chair of the department of pediatrics at the University of Connecticut School of Medicine. In 2005, he passed away at the age of eighty-seven years. A few years later, I was asked to deliver the annual Milton Markowitz Lecture in Farmington, Connecticut. With his photo on a large screen behind me, I began by telling the story of my ruptured spleen. Although I knew some of the pediatricians in the audience, none was aware that I had actually known Milton Markowitz or that he had saved my life.

The second influential event occurred as a consequence of being born with clubfeet, meaning that my feet turned awkwardly down and inward. When I was one day old, both feet were placed in casts. The left foot healed; the right foot didn't. When I was five years old, an orthopedist operated on my right foot in an attempt to straighten it. The operation should never have been performed. At the time, foot surgery was the stepchild of orthopedics. Surgeons were reluctant to operate on feet, often referring patients to podiatrists. (Clubfoot surgery would not be perfected until the late 1990s.) Nonetheless, my parents were able to find a young surgeon willing to do it. The surgery didn't go well, and I have walked with pain and a slight limp ever since.

I recovered from my clubfoot repair at Baltimore's Kernan Orthopaedics and Rehabilitation Hospital in a ward with twenty other children, all of whom had polio. Children's hospitals today feature iPads, play therapists, in-hospital television stations, and friendly therapy animals, like dogs. They also feature pullout beds so parents can sleep in their children's room. But this was 1956. No play therapists. No entertainment. No beds for parents to stay with their children. Only the fear that people might catch polio from visiting a polio ward. For that reason, access to the ward was severely restricted. My parents were allowed to visit me on Sundays from 2:00 p.m. to 3:00 p.m. only. My mother, who had experienced a medical complication from her pregnancy with my brother, never visited me. I don't remember any of the nurses befriending me. I don't remember talking to the other children. All I remember is staring out of the window next to my bed, which overlooked the front door of the hospital. I was waiting for my parents to come and rescue me. For weeks I looked out that window, hoping for something that never happened.

When I was a medical student at the University of Maryland School of Medicine in the 1970s, we rotated through Kernan's Hospital. During the rotation, I walked into the polio ward in which I had stayed as a child. It had been converted into an office. Secretaries and administrators looked up, wondering what I was doing there. The room had changed: different color, different moldings. But that window was still there. And the view from the window was the same. It was overwhelming. I'll come back to this later.

The passions created by my childhood experiences drove me into pediatrics and later into infectious diseases. The person who shaped my academic career was Dr. Stanley Plotkin.

In 1980, I began a three-year training program in pediatric infectious diseases at the Children's Hospital of Philadelphia. When I first met Dr. Plotkin, he was reading a journal published by the CDC called *Morbidity and Mortality Weekly Report*. He wanted to know the current incidence of a disease called congenital rubella syndrome.

Rubella, or German measles, was a mild infection of children, causing fever, rash, and swelling of the lymph nodes behind the ear. But when the rubella virus infected women early in their pregnancies, it could cause severe permanent birth defects of the ears, eyes, and heart. Indeed, in the early 1960s, one of every hundred pregnancies in Philadelphia was complicated by congenital rubella syndrome. It was *that* common. (Of interest, natural rubella virus infection during the first trimester of pregnancy is a known cause of autism. Which means that the MMR vaccine *prevents* one cause of autism.)

In 1979, the year before I first met him, Dr. Plotkin had invented the rubella vaccine. He was reading *Morbidity and Mortality Weekly Report* to see how his vaccine was doing. I had just finished a three-year residency program in pediatrics at the Children's Hospital of Pittsburgh, taking care of a few children at a time. While I had been doing that, Dr. Plotkin had been taking care of *entire populations of children*, both in the United States and other countries. Such was the power of vaccines. Indeed, by 2005, the rubella virus, which had caused as many as twenty thousand cases of birth defects and five thousand spontaneous abortions every year, was eliminated from the United States.

During the first year of my training, Dr. Plotkin introduced me to Dr. Fred Clark. Fred had just started a program to study rotaviruses. Fred and Stan asked me if I would be interested in working on this virus the following year. Before deciding, I met with Dr. Clark in his laboratory, which was lined with dead snakes in jars. Fred was a veterinarian who had studied snake viruses. A tall, lanky man with an ironic sense of humor, Fred explained what he hoped to accomplish, spending much of the meeting talking about the impact of rotavirus on children in the developing world and about his frequent visits to Haiti. Hanging on the wall above Fred's desk was a picture of Dag Hammarskjöld. When he saw me looking at it, Fred explained how much he admired this great statesman, a former Secretary-General to the United Nations, and one of only two people to have won the Nobel Prize posthumously. It was an interesting experience. Between the snakes, Dag Hammarskjöld, and Fred's commitment to children in the developing world, I was convinced. I signed on to the rotavirus program the next day.

In 1981, when I first I started working in Fred's laboratory, little was known about rotavirus, which had been found to cause human disease only eight years earlier. Veterinarians, on the other hand, were way ahead of the game; they had known rotavirus to be a cause of disease in animals since the 1940s. For this reason, most of the researchers in the field were veterinarians studying large animals, like pigs, sheep, and cows—all with an interest in increasing food production. Rotavirus, as it turns out, infects the young of many mammalian species. But species barriers are high; for example, cow rotavirus causes diarrhea in calves but not human infants; and human rotavirus causes diarrhea in human infants but not calves. My job was to figure out which parts of the virus made children sick and which parts evoked protective immunity. The ultimate goal was to retain the parts of the virus that elicited immunity and remove the parts that caused disease. (Which is how all vaccines are made.)

Rotavirus is fairly easy to study because its genome, the blueprint that instructs the virus how to reproduce, is segmented, which makes it easy to rearrange the genes. For example, if you take two different strains of rotavirus and infect the same cells at the same time, some genes will come from strain A, and some will come from strain B. If strain A causes diarrhea and strain B doesn't, then you can figure out which genes are the diarrhea-causing genes. Similarly, if strain A evokes antibodies that neutralize strain A but not strain B, you can figure out which genes induce protective immunity. Obviously, you can't do these studies in children. You can do them only in experimental animals. When I first started working on rotavirus, the only animal models available were large, like pigs and cows. The problem with large animals, in addition to the fact that they're expensive and that we had nowhere to put them, is that they're outbred, meaning that they vary widely in their genetic makeup. This makes it harder to study certain aspects of the immune response. The best experimental animals in which to do our studies were inbred, genetically defined mice. But, at least until this point, no one had been able to induce diarrhea in mice with strains of rotavirus that could easily be grown in the laboratory.

At the end of my first year of research, Fred, Stan, and I found a strain of rotavirus that could grow in the laboratory and caused diarrhea in baby mice. I remember the day. After gently palpating the mouse's abdomen, a thin stream of liquid diarrhea emerged. This was great. Now we had a small-animal model to answer the questions we needed to answer. At least theoretically, we should have been able to create strains of rotavirus that didn't cause diarrhea but did evoke protective antibodies. First, we had to find an animal strain that didn't cause diarrhea in infants. To do this, we drove to the large-animal facility at the University of Pennsylvania School of Veterinary Medicine in Kennett Square, Pennsylvania, to collect diarrhea from calves that were sick with rotavirus. Then we collected diarrhea from infants infected with rotavirus. For the next ten years, we made combination viruses between these cow and human strains and, after studying what amounted to thousands and thousands of mice, created five combination viruses that we thought could be a vaccine for children. (It's depressing to me that I can actually summarize ten years of work in one paragraph.)

During the years we did these studies, our mice were housed in the Wistar Institute, located on the campus of the University of Pennsylvania in Philadelphia. On weekends, when my wife worked, I brought the kids with me. One day, while I was working with the mice, Will, who was about three years old, was petting a baby mouse in the corner. Baby mice are about the size of chickpeas. Just as I looked up, he was putting one into his mouth. Fortunately, I stopped him before he ate it. (This would have played out well at home. "Hi, honey. I'm home. Will had a great time with the mice today. I think he really loves them. Oh, by the way, he ate one.")

Fred, Stan, and I were hopeful that what we had found in mice would also be true in children. But we also knew that this might not be the case. "Mice lie and monkeys exaggerate," said the Wistar Institute vaccine researcher David Weiner. In 1988, we submitted a patent application for our vaccine to the Patent and Trademark Office in Crystal City, Virginia. Stan had explained that no pharmaceutical company would advance our vaccine unless the technology was protected.

I don't want to create the impression that we were alone in doing these studies. Researchers at Baylor University, the Children's Hospital of Cincinnati, the National Institutes of Health, Stanford University, and research laboratories in Australia, England, France, Germany, Italy, Portugal, and Spain had also been studying rotavirus and finding the same things we had found. Indeed, one of the many wonderful aspects of science was getting to spend time with these other researchers, all of whom were invariably excited to talk about their work and share their findings—an international brotherhood and sisterhood.

In 1988, Dr. Plotkin left the Children's Hospital of Philadelphia, leaving Fred and me to take the next step on our own. We had to find a pharmaceutical company willing to see if what we thought was a rotavirus vaccine actually was a rotavirus vaccine. Merck was the most interested. For the next sixteen years, Merck scientists performed studies in adults, then older children, then younger children, and then infants, proving that all five strains of our cow–human rotaviruses needed to be in the vaccine (these were called proof-of-concept studies); that we had inoculated children with the right amount of each vaccine strain (dose-ranging studies); that the vaccine viruses didn't break down (real-time stability studies); and that, by creating the right vial, the vaccine could easily be squirted into the mouths of babies. These studies, which cost a little more than $1 billion in total, ended with a four-year prospective, placebo-controlled trial that took place in eleven countries and studied more than seventy thousand children, and which cost about $350 million. This so-called Phase III study showed that the vaccine worked and was safe. Nonetheless, during the sixteen years that it took to do this research of development, we faced many "go–no go" decisions. Could the vaccine strains be mass-produced efficiently? Would the vaccine have a shelf life long enough to effectively immunize children in the United States and around the world? And on and on. The vaccine, and we, died a thousand deaths.

After the results of the Phase III trial had been tabulated, I was satisfied that what we had found in mice had been confirmed in children. The vaccine appeared to be safe and effective, at least as far as it had been tested. I was, however, nervous about the next step: giving the vaccine to tens of millions of children in the United States and other countries. For several reasons, these concerns were well founded. An earlier rotavirus vaccine, made by combining a simian rotavirus with human rotaviruses, had been used in the United States between 1998 and 1999. The vaccine, developed by researchers at the National Institutes of Health in collaboration with Wyeth, had been found to be a rare cause of intestinal blockage. Called intussusception, this blockage occurs when the small intestine telescopes into itself and gets stuck. One child died as a consequence. This problem was found only *after* the vaccine had been licensed and recommended for all infants. Also, while studies performed in tens of thousands of children pre-licensure can rule out uncommon side effects, they can't rule out rare side effects. As Dr. Maurice Hilleman, the Merck researcher who developed nine of the fourteen vaccines currently given to infants and young children, famously said, "I never breathe a sigh of relief until the first three million doses are out there."

The notion that we could have the same problem as the previous rotavirus vaccine was frightening. As our vaccine got closer to licensure, Fred and I became obsessed with the fact that we might be missing something. Was it possible that one of the rotavirus proteins in our vaccine mimicked other proteins in children, like ones that sit on the surface of pancreatic cells that make insulin? If that were the case, our vaccine could cause diabetes. Or, if rotavirus proteins mimicked proteins on the surface of cells that line joints, wouldn't it be possible that our vaccine could cause arthritis? Fred and I scanned gene databases looking for any similarities between rotavirus proteins and human proteins. Although one would imagine that a seventy-thousand-person trial that showed that the vaccine was safe would bring relief, it didn't. The previous rotavirus vaccine had been distributed for about ten months and given to a million children before scientists found the problem, which occurred in about one of every ten thousand children inoculated. Would ours be any different? "When the gods are angry, they grant your wish," warns an ancient Chinese proverb. We had been granted our wish.

In February 2006, the Food and Drug Administration licensed our vaccine, called RotaTeq, for use in the United States. That same month, the CDC recommended that children receive the vaccine at two, four, and six months of age. Since licensure, the number of children in the United States hospitalized for dehydration caused by rotaviruses has declined dramatically. At the Children's Hospital of Philadelphia, during the winter months, we would typically admit about four hundred children with dehydration caused by rotavirus. Now it is rare for us to admit any child to the hospital with rotavirus infection. Indeed, some young pediatricians have never seen a case of rotavirus, a disease that had previously affected about four million children in the United States every year. In 2013, the World Health Organization recommended that rotavirus vaccines be included in all childhood immunization programs. Now, in Africa, Asia, and Latin America, rotavirus vaccines are saving hundreds of lives a day.

RotaTeq is the professional accomplishment of which I am most proud. But I was lucky. I was in the right place at the right time. Stan Plotkin, the inventor of the rubella vaccine, the principal investigator on the clinical trials of the modern rabies vaccine, and the senior editor of the definitive textbook on vaccines, is the single most respected, most accomplished vaccine researcher in the world. I was fortunate to train with him. Fred Clark, with his background in veterinary medicine, was the perfect person to work on a vaccine that included both human and animal strains. Also, Fred had an uncanny sense of judgment. When things went wrong in the lab, he invariably found the problem. And Fred was enormously patient. I had had no laboratory experience before working with him. But he calmly taught me how to be successful. When I wrote my first paper, he patiently crossed out every sentence and rewrote it. (I'm not kidding. Every sentence.) As I got better at writing—which is to say when I finally learned to write like Fred—he would cross out fewer and fewer sentences. Stan wasn't quite as patient. While reading my first paper, he started to cross out every sentence, eventually giving up and writing, "Read Strunk and White's *Elements of Style*!"

I've received many accolades for my work on the rotavirus vaccine. I've been elected to the National Academy of Medicine, honored by Bill and Melinda Gates as part of their Living Proof Project, and elected to the American Academy of Arts and Sciences in the same class as the Nobel Prize winners Holland Cotter and Brian Kobilka, the astrophysicist Neil deGrasse Tyson, the novelist Tom Wolfe, the Tony Award winner Audra McDonald, the Nike cofounder Phil Knight, and the singer-songwriter Judy Collins. The point being, I'm overrated. I'm not saying that I didn't work hard or that I didn't do the doable. I'm just saying that anyone could have stepped into the situation with Stan Plotkin and Fred Clark and do what I did. So whenever I'm asked to receive these honors, I feel like they're undeserved. At the very least, Stan and Fred should always be standing there next to me. And in a sense they always are. (Fred passed away in April 2012.)

· · · ·

AFTER SPENDING TWENTY-SIX YEARS WORRYING ABOUT GETTING papers published and grants funded, worrying about whether what had worked in mice would also work in children, and worrying about what would happen after our vaccine had been given to tens of millions of infants, the project was over. What happened next was something that I hadn't thought about, but which would become—at least in the minds of those who vilify me—my defining characteristic. I became financially secure. Although Fred, Stan, and I owned the patent on the rotavirus vaccine, the Children's Hospital of Philadelphia (CHOP) and

the Wistar Institute owned the rights to any intellectual prop-
erty we created or developed. So, for all practical purposes,
CHOP and Wistar owned the patents. Therefore, when CHOP
sold out its rights in 2008, Stan, Fred, and I were subject to
CHOP's patent policy.

Because I am the inventor of a patented medical product, the
notion that I have been unduly influenced by the pharmaceutical
industry will no doubt follow me for the rest of my life. There is
simply no getting around it. To anti-vaccine activists and certain
members of the media, I will always be the poster child for conflict of interest—even though any compensation that I received
for co-inventing a vaccine ended in 2008. According to them,
I am being paid by pharmaceutical companies to say that vaccines don't cause autism or a variety of other illnesses. I am the
lowest form of human being: someone willing to sacrifice children for personal gain. There are, however, a few problems with
this logic.

First, apart from the fact that I don't have a financial tie to
any pharmaceutical company, accusations of conflicts of interest
would make a lot more sense if scientific studies didn't consistently support my position. Seventeen studies have shown that
the MMR vaccine doesn't cause autism; seven studies have
shown that thimerosal-containing vaccines don't cause autism;
and two studies have shown that the many vaccines given in
infancy don't cause autism. The reason that I say that vaccines
don't cause autism is that vaccines don't cause autism. Further,
many other studies have shown that vaccines don't cause the
wealth of other problems claimed by anti-vaccine activists.
I wasn't an author on any of those studies. I'm simply reporting
what scientific studies have shown to be true. Although anti-
vaccine activists argue that I am a Pharma shill in the same way
that people shilled for the tobacco industry, there is one critical
difference. Cigarette smoking *does* cause lung cancer. Vaccines,
on the other hand, don't cause the diseases claimed.

Second, the logic of my alleged nefarious alliance with the
pharmaceutical industry is at best tortured. Let me see if I've got

this right. I worked for twenty-six years to make a vaccine that helps children so that I could make money so that I could lie about vaccines and hurt children. Does that make sense? Sadly, at least to the hard-core anti-vaccine activists, it does.

Third, how should I have handled this differently? Assuming that preventing hundreds of thousands of deaths in children every year is a good thing, that only pharmaceutical companies have the resources and expertise to make a vaccine, and that the only way that companies will make a vaccine is if the technology is protected (meaning patented), then how should I have proceeded? Should I have just been satisfied with publishing papers about rotaviruses and getting grants and left it at that? And why should I disqualify myself from the public debate about vaccines just because I am the co-inventor of one? It seems to me, given that I have participated in shepherding a vaccine from early research into a medical product, have published almost two hundred papers in the field of viral immunology, have received a series of competitive grants from the National Institutes of Health, am the co-editor of the definitive textbook on vaccines, and have published five medical narratives about vaccines, I am a reasonable person to educate the public on the subject.

Finally, if I'm a shill for the pharmaceutical industry, I haven't been very good at it. When I headed the CDC working group on polio vaccines, I pushed to eliminate the oral polio vaccine in the United States; as a consequence, the makers of that vaccine lost millions of dollars. Also, I was a lone voice against the use of the smallpox vaccine in the early 2000s, going on *60 Minutes* and *PBS NewsHour* with Jim Lehrer to make my case. In 2016, I made a video for the website *Medscape* decrying the ineffectiveness of the nasal spray influenza vaccine FluMist, which at the time was also eliminated for use in the United States, costing the company tens of millions of dollars a year. The point being, I'm not a vaccine cheerleader. I'm a pro-science advocate, wherever the science falls.

Although I would argue that I have never had a real conflict of interest, the minute the media asks you if you have any connection to a pharmaceutical company, you've lost. Transparency has morphed into bias.

Ed Asner (who starred in *The Mary Tyler Moore Show* and *Lou Grant*) tells a story about John Wayne when they were filming the movie *El Dorado*. Wayne asked, "Where is that New York actor?" which Asner took as a mildly anti-Semitic comment. Then, in one scene, Asner throws a saddlebag at Wayne in such a way that he can only look awkward catching it. When members of the media ask me about my association with a pharmaceutical company, it's like Ed Asner throwing that saddlebag at me. I can only look awkward answering. Almost anything I say will sound defensive. I've said, "Only pharmaceutical companies have the resources and expertise to make vaccines. What other choice did I have?" (Which sounds like I'm agreeing that vaccine makers are evil.) I've said, "Were we supposed to make tens of millions of doses of the vaccine in our lab?" (Which is too glib.) I've said, "I'm proud of our association with Merck. The scientists there were outstanding, and we made something that has clearly benefited children." (Which, although true, sounds like a statement written for me by the company.) The point being, there is no winning.

The important thing for scientists who are asked this unanswerable question is to know that they have done the right thing for the right reasons. This, I'm afraid, will be your only solace.

Among the celebrities and politicians who have attacked me as a shill for the pharmaceutical industry, no one has been more persistent, more mean-spirited, or apparently better funded than Robert F. Kennedy Jr. I've never met RFK Jr., but I talked to him by phone in 2005.

Kennedy left a message on my voice mail. He wanted to talk about vaccines; specifically, he wanted to know more about thimerosal, the ethylmercury-containing preservative. Later, he sent me an email signed "Bobby." This was completely cool. One of Robert F. Kennedy's children had called me. His father, RFK, was a hero to me. I had read his book *Thirteen Days*, about the Cuban Missile Crisis and loved it. In fact, I often recited a statement that RFK had paraphrased from Dante: "The lowest reaches in hell are reserved for those who, in times of moral crisis, are ambivalent." That quote was a call to arms. Stand up for what you believe in. I couldn't wait to talk to Mr. Kennedy's son on the phone. This was going to be great. (If you're thinking that this is another I'm-an-idiot story, you're right.)

I talked with RFK Jr. for a long time. He said that several parents had recently approached him worried about mercury in vaccines. He wanted to know whether their concerns were valid. He gave me plenty of time to answer his questions about mercury. He also wanted to discuss how vaccines work and how they're made. I talked about my work on a rotavirus vaccine. And I talked about how much I had admired his father. He seemed amiable and accepting. After I hung up, I thought I had done a pretty good job.

In July 2005, *Rolling Stone* magazine published an article by Robert F. Kennedy Jr. It was titled "Deadly Immunity." Kennedy had sandbagged me. He wasn't interested in learning about thimerosal. He just wanted to paint me as yet another industry insider who didn't care about children—someone who was trying to explain away thimerosal as harmless when Kennedy believed the opposite to be true. He wrote that I had defended thimerosal because, as the co-inventor of a rotavirus vaccine, I stood to make money off of it. The reason, he said, was obvious: The rotavirus vaccine was "laced with thimerosal." (No rotavirus vaccine has ever contained thimerosal, including the one we developed at the Children's Hospital of Philadelphia.) The article also stated that "by the age of six months, [babies] were being injected with levels of ethylmercury that were 187 times greater than the EPA's [Environmental Protection Agency's] limit for daily exposure to methylmercury." (Children didn't receive 187 times the EPA limit for methylmercury; they received 0.4 times the limit. Or, said another way, were well within the EPA limit.) The point of Kennedy's article was that doctors and public health officials knew that thimerosal was harming children but had chosen to cover it up—all part of a massive conspiracy to hide the truth. Doctors like me didn't care about children. All we cared about was our own pockets, children be damned.

After I read "Deadly Immunity," I called an editor of *Rolling Stone* magazine. I wanted to know why he hadn't fact-checked the article. But he just kept saying to me, "We stand by our story." I couldn't believe it. I asked him to just look at the package inserts of the various vaccines to learn about the quantities of thimerosal, to look at federal guidelines that showed that Kennedy was off on his calculations, and to look at the clinical trials of the rotavirus vaccine (which had been published in the *New England Journal of Medicine*) to see that the vaccine didn't contain thimerosal, as Kennedy had claimed. I told him that no live weakened viral vaccines have ever contained thimerosal. But he just kept saying, "We stand by our story." (I suspect that the *Rolling Stone* editor was trying to channel Ben Bradlee, the executive editor of the *Washington Post* who had said exactly the same thing in the movie *All the President's Men*. Bradlee had been standing behind the stories of the Watergate cover-up written by Bob Woodward and Carl Bernstein. The difference between Ben Bradlee and the *Rolling Stone* editor was that there actually *had been* a cover-up of the Watergate break-in. The *Rolling Stone* editor, on the other hand, was defending something that was not only wrong, but could have been shown to be wrong in a matter of minutes.) I told the editor that these were facts that could easily be checked. I got angrier and angrier until he started to call me "Sir" (never good). Then he threatened to hang up.

Before Kennedy published "Deadly Immunity" in *Rolling Stone*, it was published on Salon.com. However, just days later, the editors at *Salon* realized the story was wildly inaccurate and amended it to correct several factual errors, which the editor-in-chief, Kerry Lauerman, said "went far to undermine Kennedy's exposé." Continued criticism of the story "further eroded any faith we [at Salon] had in the story's value," however, and the piece was removed from the *Salon* website with an apology to readers.

Robert F. Kennedy Jr. isn't alone in using ad hominem attacks to lessen the credibility of those who have stood up for science.

Although several members of the media have been willing to paint pro-science activists as shills for industry, one instance that stands out is Sharyl Attkisson's vilification of me, the American Academy of Pediatrics, and the wonderful vaccine advocacy group, Every Child By Two. Attkisson's segment, "How Independent Are Vaccine Makers?," aired on *CBS Evening News* on July 25, 2008. The show began with a teaser: "For years, some parents and scientists have raised concerns about vaccine safety, including a possible link to autism and ADD [attention deficit disorder]. Many independent experts sided with government officials and other scientists who say there's no possible connection. But how 'independent' are they? *CBS News* investigative correspondent Sharyl Attkisson shares what she found."

Attkisson told her viewers that both Every Child By Two and the American Academy of Pediatrics had received funds from pharmaceutical companies. Regarding my illicit behaviors, she said, "Then there's Paul Offit, perhaps the most widely quoted defender of vaccine safety. He's gone so far as to say babies can tolerate ten thousand vaccines at once." (Lord, I wish I'd never said that.) "Offit was not willing to be interviewed on this subject, but like others in the *CBS News* investigation, he has strong industry ties. In fact, he's a vaccine industry insider. Offit holds a $1.5 million research chair at Children's Hospital, funded by Merck. He holds the patent on an anti-diarrhea vaccine he developed with Merck, RotaTeq, which has prevented thousands of hospitalizations. And future royalties for the vaccine were just sold. . . . Dr. Offit's share of vaccine profits? Unknown." (For the record, the inventors' share of my hospital's royalties is publicly available.)

Attkisson then implied that I was unwilling to admit my shameful association. "This is how Offit described himself in a previous interview," she said, with a figurative raised eyebrow. "I'm the chief of infectious diseases at Children's Hospital of Philadelphia and a professor of pediatrics at Penn's medical school," I said.

Neither the American Academy of Pediatrics nor Every Child By Two has ever hidden the fact that they have received unrestricted grants from pharmaceutical companies to educate the public about vaccines. The goals of both groups are clear. The better educated the public is about vaccines, the more likely they will be to vaccinate their children, and the more likely it is that those children won't suffer diseases that could hurt or kill them. Attkisson never showed how these unrestricted educational grants had caused either the American Academy of Pediatrics or Every Child By Two to say something that wasn't true. Further, I have never denied the fact that a pharmaceutical company made our vaccine or that I was awarded an endowed chair given by Merck to the Perelman School of Medicine at the University of Pennsylvania in the name of the vaccine researcher Maurice Hilleman. Not surprisingly, I'm proud of this.

Regarding her takedown of me, Attkisson made two mistakes: one out of ignorance, the other that I believe bordered on journalistic malpractice. Endowed chairs are analogous to unrestricted educational grants. As the recipient of the Hilleman endowment, I don't have to report to Merck about how the money is spent. I have to report only to the Children's Hospital of Philadelphia and the University of Pennsylvania School of Medicine. Indeed, I don't have to spend the money on anything related to vaccines. Endowments, which abound in academic medicine, allow for academic freedom. For example, academics working at the Mattel Children's Hospital in Los Angeles aren't expected to defend plasticizers in children's toys any more than those who work at the Icahn School of Medicine in New York City are expected to support hostile takeovers or those who work at the David Geffen School of Medicine at UCLA are expected to promote movies like *Beetlejuice* or *Little Shop of Horrors*.

The aspect of Attkisson's story that I felt had crossed the line from bad reporting to misrepresentation was the quote she had used from me. As she stated correctly in her report, I had refused to be interviewed, as had representatives from the American Academy of Pediatrics and Every Child By Two. We knew that this was going to be a hit piece. The way the segment was framed, Attkisson makes it sound that by simply stating my name and where I worked, I was denying any association with a pharmaceutical company. But I was never asked to respond on camera to what Attkisson had claimed were my illicit associations. The clip of me was taken months earlier. During that particular interview, the producer asked me to state my name and where I worked. So I stated my name and where I worked. Years later, when people from *CBS Evening News* came to my office to interview me about a measles outbreak, I told the producer what had happened during the Attkisson piece. She apologized and said that those were darker days. Attkisson currently hosts a public affairs program called *Full Measure with Sharyl Attkisson*, which is operated by the conservative Sinclair Broadcast Group.

I never told my parents about Attkisson's *CBS News* segment before it aired, hoping that they wouldn't see it. But they did. My mother called me later that night. She was thrilled. "I thought you looked good," she said. My mother has a very low bar for what is considered a successful television appearance.

Because Sharyl Attkisson was a known anti-vaccine journalist, I was prepared for her. The television interview that I wasn't prepared for was with Matt Lauer on *Dateline NBC*. Lauer has a disarming, friendly manner; he doesn't appear to be someone with an agenda or someone who is going to ambush his guest. So when I agreed to appear on *Dateline NBC* to talk about the MMR–autism controversy, I thought we would be talking about the studies that had exonerated the vaccine. But I was wrong. Lauer wanted to talk about me.

"You have been called some horrible things," said Lauer. "They have called you a baby killer. They've said you have blood on your hands."

"It's been pretty brutal," I said. "But . . . I think that if they knew me, they wouldn't feel that way."

"You have had threats made against you," said Lauer, "legitimate threats."

"That's right," I said, alluding to an episode in which a telephone caller indirectly threatened my children by revealing that he knew where they went to school. "That was the one time I considered stopping all of this."

Lauer pressed on: "Some of the people who believe Dr. Wakefield look at you, and they say, 'Wait a second . . . here's a guy who created a vaccine, so of course he's not going to say anything that would call into question other vaccines; it's just against his DNA.' "

"I worked for twenty-five years to create a vaccine that has the capacity to save two thousand lives a day," I said. "What motivated me to do that and the reward from doing that was of course not financial. My interest [now] is trying to represent the science of vaccine safety so that parents can understand it so that they won't make bad decisions that put their children at risk."

Matt Lauer wanted me to admit that, at the very least, given that I was the co-inventor of a vaccine—and that I had been compensated for that invention—I could understand how people could hate me. But the truth is that I don't understand it. I don't understand how anyone can believe that there is a conspiracy to hurt their children with vaccines and that people like me are part of it.

Given the level of anger that I seem to evoke, friends and coworkers have asked me why I persist in trying to educate the public about vaccines. Why do I continue to subject myself to so much anger? I give two answers, both of which are accurate but incomplete. I say that I do it because every year children suffer and die from vaccine-preventable diseases, invariably because their parents had chosen not to vaccinate them. (That's the moral-high-ground answer.) Or I say that I can't abide professional anti-vaccine groups, like Generation Rescue, Moms Against Mercury, the National Vaccine Information Center, and SafeMinds, because they are often lawyer-backed and politically connected. That I see them as evil—a sort of cottage industry to hurt children—and that they shouldn't be allowed to get away with it. (That's the moral-low-ground answer.)

But I think the real answer lies somewhere else. And it's far more selfish. Sometimes when I give talks, I find myself fighting back tears when I tell stories of children suffering from preventable diseases or from a parent's misplaced reliance on alternative remedies or from faith healers who refuse children medical care. Indeed, all the books I've written are about child advocacy. Where is all of this emotion coming from? I think that it might have something to do with seeing myself in all of these children—seeing in every story a little boy staring out of that hospital window. A little boy who was vulnerable and helpless and alone.

But what do I know? I'm not a psychiatrist. Maybe I just hate to see the bad guys win.

CHAPTER 12

A Ray of Hope

History, despite its wrenching pain, cannot be unlived,
but if faced with courage, need not be lived again.

—MAYA ANGELOU

In the world of science advocacy, victories are often hard earned and incomplete. No matter how compelling, dramatic, and clear the messaging, science denialists are not going to put down their signs decrying their fears, stop their protests, and say, "You know, I never thought about it that way. Now that I look carefully at the studies, I see that you're right. And I want to take a moment to tell you just how much I appreciate the academic community's willingness to spend millions of dollars and thousands of hours to answer our concerns. Thanks for all you do. Really. Thanks." If this ever happens, then, as predicted in the Book of Isaiah, "Men will beat swords into ploughshares and their spears into pruning hooks; nations will not lift up sword against other nations, nor will they learn war anymore." Scientists and science denialists will stand shoulder to shoulder on large green meadows, swaying back and forth, smiling and singing in perfect harmony about apple trees and honey bees and furnishing the world with love—like that Coke commercial at the end of *Mad Men*.

Although I think this scenario is unlikely, the situation has definitely improved.

. . . .

ON MARCH 21, 2016, THE LEGENDARY ACTOR ROBERT DE NIRO announced that he would premiere *Vaxxed: From Cover-Up to Catastrophe* at his prestigious Tribeca Film Festival, which he cofounded in 2002. *Vaxxed* was written and directed by Andrew Wakefield. The festival's website provided a glimpse of what was to come:

> The most vitriolic debate in medical history takes a dramatic turn when senior-scientist-turned-whistleblower Dr. William Thompson for the Centers for Disease Control turns over secret documents, data, and internal emails confirming what millions of devastated parents and "discredited" doctors have long suspected . . . what's behind the skyrocketing increase of autism.

Since Andrew Wakefield had published his study in *The Lancet*, seventeen studies had shown that he was wrong: MMR didn't cause autism. But now, at least according to *Vaxxed*, it appeared that Andrew Wakefield had been right all along. The only reason that these other studies had dismissed his hypothesis was that they had been falsified. Wakefield hadn't been a fraud; university researchers and public health officials around the world had been the frauds. And now, finally, a government scientist was willing to stand up and blow the whistle.

"Wow," says Wakefield in the trailer for *Vaxxed*. "The CDC had known all along there was the MMR–autism risk." Andrew Wakefield, apparently, was back.

Thirteen years earlier, in 2003, Wakefield had been the star of the docudrama *Hear the Silence*. Now he was the director and star of a movie that would exonerate him. Robert De Niro announced that *Vaxxed* would premier on April 24, 2016, the last day of the festival.

The media's reaction to De Niro's announcement was swift and one-sided.

Michael Specter, a staff writer for the *New Yorker*, wrote, "It's shocking. This is a criminal who is responsible for people dying. This isn't someone with a 'point of view.' It's comparable to Leni Riefenstahl making a movie about the Third Reich, or Mike Tyson making a movie about violence toward women. The fact that a respectable organization like the Tribeca Film Festival is giving Andrew Wakefield a platform is a disgraceful thing to do."

Michael Hiltzik, in the *Los Angeles Times*, wrote, "The damage done to public health by the British ex-physician Andrew Wakefield . . . has been incalculable. Wakefield's claims have been conclusively discredited everywhere but in the fever swamp of the anti-vaccine movement—and now in the glamorous environment of the Tribeca Film Festival. How on earth did this documentary full of anti-vaccine lies get into Tribeca? The answer may have much to do with Hollywood's taste for anything that promotes drama and controversy, no matter how irresponsible."

Alexandra Sowa, in the *New York Daily News*, wrote, "I am in complete agreement that we need to discuss autism, a serious disease that now affects one in sixty-eight U.S. children. But there is no longer room for Andrew Wakefield or vaccines to be part of the conversation. . . . For many, he still serves as a beacon of hope in the fight against autism. But do not be fooled: He is a snake oil salesman and Robert De Niro is helping him peddle his dangerous wares."

All this negative press, and the movie hadn't been released yet.

In the end, *Vaxxed* didn't resurrect the career of Andrew Wakefield. And it didn't start a revolution. Colin McRoberts, in an article titled, "*Vaxxed*: From Cover-Up to Catastrophe to Cancellation to Insignificance," accurately predicted its impact: "[*Vaxxed*] will be seen as incredibly important to a small minority of anti-vaxers who would have been anti-vaxers anyway, as a noxious piece of harmful propaganda to actual experts, and as a temporary curiosity to the mainstream."

Following its theatrical release, *Vaxxed* was ignored by newspapers, magazines, and trade journals across the country. The few who reviewed it were brutal.

On April 19, 1982, nearly thirty-five years before the release of *Vaxxed*, Lea Thompson, a veteran correspondent for NBC News, aired a one-hour special titled "DPT: Vaccine Roulette." (The vaccine, which is actually called DTP, is a combination vaccination against diphtheria, pertussis [whooping cough], and tetanus.) The documentary claimed that the pertussis component of the DTP vaccine caused permanent brain damage. The images were riveting. Children stared vacantly into space with withered arms and legs, seizing, crying, helpless. Parents all told the same story; our children were fine; then they got this vaccine, and look what happened.

The media covered "Vaccine Roulette" as fact. Anti-vaccine advocacy groups were born. Congress held special hearings to determine whether vaccines were doing more harm than good. Manufacturers abandoned vaccines following an avalanche of lawsuits. Massive outbreaks of pertussis swept across the country. Although dozens of studies later showed that the pertussis vaccine hadn't caused brain damage, the notion that vaccines carried hidden dangers was born.

Following the airing of "DPT: Vaccine Roulette," the anti-vaccine movement was riding high. When the media wanted to get parents' perspectives on vaccines, they called Barbara Loe Fisher, the founder of the anti-vaccine group *Dissatisfied Parents Together* (derived from the purported problem with the DTP vaccine), which would later become the National Vaccine Information Center. Fisher was everywhere, a one-stop shop for the media. When the CDC recommended the *Haemophilus influenzae* type b (Hib) vaccine for infants in the late 1980s, Fisher appeared on a national television program warning parents that the vaccine caused diabetes. When the hepatitis B vaccine was recommended for all newborns in 1991, Barbara Loe Fisher worked with ABC's Sylvia Chase on a *20/20* episode claiming that the vaccine caused chronic arthritis, multiple sclerosis, and sudden infant death syndrome. When the pneumococcal vaccine was recommended for infants in 2000, Fisher again appeared on national television warning against its routine use. Parents who listened to Barbara Loe Fisher's misleading claims put children at unnecessary risk of meningitis, pneumonia, bloodstream infections, hepatitis, cirrhosis, and liver cancer.

Fisher wasn't the only anti-vaccine activist with ready access to the media. J. B. Handley, the cofounder of Generation Rescue, Jenny McCarthy's autism research organization, was often quoted in national newspapers and magazines and appeared on television shows and in documentaries. Handley believed that vaccines had caused his son's autism and that by injecting children with drugs that bind mercury, children with autism could be cured. He traveled around the country to preach his ill-founded beliefs, calling his recruits "rescue angels." Other anti-vaccine activists, like Sallie Bernard and Lyn Redwood of SafeMinds and Mark Blaxill of Age of Autism, were often given access to the media. At the center of it all, however, was Andrew Wakefield.

For years, these groups held sway with the media. Then, just as quickly as it had come, all of this mainstream media attention waned. Today, reputable news sources rarely give a voice to the scions of the anti-vaccine movement. Quite the opposite. Now we hear an avalanche of voices from the other side. Celebrities like Marc Anthony, Kristen Bell, Julie Bowen, Jennifer Garner, Sarah Michelle Gellar, Salma Hayek, Kim Kardashian, Jennifer Lopez, Amanda Peet, Keri Russell, and Marissa Winokur; television journalists like Richard Besser, Campbell Brown, Erin Burnett, Anderson Cooper, Michael Smerconish, and Nancy Snyderman; radio talk show hosts like Dom Giordano and Margaret Hoover; athletes like Kristi Yamaguchi and Deion Branch (who made eleven receptions for the New England Patriots in Super Bowl XXXIX, single-handedly defeating my beloved Philadelphia Eagles but who can now be forgiven); cultural and political icons like Bill Gates, Michelle Obama, and Mark Zuckerberg; comedians like Samantha Bee, Stephen Colbert, John Oliver, Penn and Teller, and Jon Stewart; science bloggers like Matt Carey, David Gorski, David McRaney, Colin McRoberts, Steven Novella, Ken Reibel, and Michael Simpson; and journalists like Arthur Allen, Gardiner Harris, Ron Lin, Anita Manning, Maryn McKenna, Donald McNeil Jr., Seth Mnookin, Michael Specter, Mike Stobbe, Liz Szabo, and Trine Tsouderos, among many others, have all stepped forward not only in defense of vaccines, but in angry opposition to the anti-vaccine movement.

What happened? Why has the media largely abandoned its mantra of balance in favor of the science that has exonerated vaccines?

One possible explanation is that the science has matured. When Andrew Wakefield claimed that the MMR vaccine caused autism in 1998, no one could point to a study showing that he was wrong. Since then, seventeen studies have been published. Now journalists can confidently stand on a mountain of evidence. But the same could be said for climate change, which also stands on a mountain of evidence but is still often presented by the media as a controversy. Indeed, whole political parties continue to deny its existence. The availability of clear, overwhelming scientific evidence alone doesn't appear to be enough to explain this sea change against the anti-vaccine movement.

Another possible explanation is that the press and the public have finally tired of conspiracy theories. The premise of *Vaxxed* is that the government and pharmaceutical companies have joined forces to defeat Andrew Wakefield, who casts himself as a counter-cultural hero. At the time of the release of *Vaxxed*, I was asked to write a review of the movie for the *Hollywood Reporter*, in which I concluded, "For people who believe that President Barack Obama is not a U.S. citizen, that the moon landing was filmed on a Hollywood sound stage, and that an intergalactic board of elves and fairies are trying to get the IRS out of Puerto Rico, this movie is for you." Judging from the number of hate-filled emails and phone calls I received after writing this review, conspiracy theories and conspiracy theorists are alive and well. So I don't think that's it.

Or maybe the anti-vaccine movement has been a victim of its own success. Now—scared by the notion that vaccines might be causing chronic disorders—a critical number of parents have chosen not to vaccinate their children. As a consequence, measles outbreaks occurred in the United States in 2014, 2015, and 2017. Nothing educates like highly contagious viruses. The combination of the recent measles outbreaks, the fraudulent study that promoted the MMR vaccine as a cause of autism, the studies that exonerated the MMR vaccine, and the vigorous response by clinicians, academics, and science bloggers in support of those studies, at least in part account for the marginalization of the anti-vaccine movement by the press. Still, I don't think this explains everything. Rather, I think that the dramatic fall of the anti-vaccine movement can be attributed in large part to one man: Andrew Wakefield, one of the greatest cautionary tales of the modern era.

When Wakefield stepped forward in 1998 to tell the world that the MMR vaccine caused autism, the media embraced him. Thousands of articles were written about his discovery, most claiming that at the very least he had raised an important question. Wakefield appeared on virtually every morning and evening television program; Ed Bradley interviewed him on *60 Minutes*, and Representative Dan Burton asked him to speak in front of an important congressional committee. Wakefield was the newly minted hero of the anti-vaccine movement—an accomplished physician who provided an explanation for something that had previously been unexplainable.

Then Wakefield was found not only to be wrong, but to have acted fraudulently. And nobody likes a fraud. We learned, primarily from the investigative efforts of Brian Deer, that Wakefield had misrepresented clinical and biological data; that he had received hundreds of thousands of dollars to support research on behalf of parents in the midst of suing pharmaceutical companies; that he had filed a patent on what he believed would be a safer measles vaccine; and that he had drawn blood from children at his young son's birthday party. For these and other transgressions, Wakefield lost his license to practice medicine, and his now infamous *Lancet* paper was retracted. Andrew Wakefield had become a pariah, an icon for bad science: his "study" appearing on biology exams as an example of what scientists shouldn't do.

But Wakefield didn't give up. With Polly Tommey, Wakefield helped create the Autism Media Channel, a video production company that portrays children with autism as vaccine-damaged, further stigmatizing them. During the promotion of *Vaxxed*, Andrew Wakefield, Polly Tommey, and Del Bigtree (the producer) traveled from state to state, venue to venue. Crowds were small and the media generally absent. Although Wakefield had always believed that pharmaceutical companies and public health agencies had conspired to defeat him, he had seriously miscalculated how unpopular he had become. Journalists, child activists, and parents of children with autism now saw him not only as a pariah, but as a threat to children with autism. Ken Reibel, a parent of a child with autism and the curator of the blog *Autism News Beat*, said, "He's still trying to tell the world he's not the greatest scientific fraud of the twentieth century. He's scared that is what he's going to be remembered for." Matt Carey, a blogger, physicist, and another father of a child with autism, wrote, "It's important to note that Wakefield is NOT part of the autism community. He's a parasite who latched on to our community twenty years ago and now we are his main source of income. He is a problem, not a solution." At the time of the release of *Vaxxed*, 85 percent of parents of children with autism didn't believe that vaccines had been the cause. In his many interviews, Wakefield had consistently repeated the theme that "we need to listen to what the parents are telling us." He wasn't listening anymore.

The anti-vaccine movement, by attaching itself to Andrew Wakefield's rising star, has now fallen with him. Colin McRoberts, a journalist who had interviewed Wakefield during the "Conspira-Sea Cruise," summed it up best: "Here's the big takeaway for the anti-vaccine movement. Andrew Wakefield is poisonous. He is death to your credibility. I am actually happy that anti-vaxers won't listen to me here, because I don't know how to cut the ethical Gordian Knot; on the one hand, I want Wakefield involved in all high-profile anti-vax efforts because he sabotages them with his very presence. . . . On the other hand, I also think it's long past time for Wakefield to disappear into ignominious retirement."

Although it might seem counterintuitive, when communicating science and health information to the public, it can be a godsend to have someone like Andrew Wakefield on the other side.

EPILOGUE: THE END
OF THE TOUR

*Though passion may have strained, it must not break
our bonds of affection. The mystic chords of memory
will swell when again touched, as surely they will be,
by the better angels of our nature.*

—Abraham Lincoln

Although science is under siege, science advocates are fighting back.

On April 22, 2017, largely in response to the Trump administration's anti-science positions, concerned citizens marched on behalf of science. In four hundred cities in the United States and six hundred cities around the world, "from the Washington Monument to Germany's Brandenburg Gate and even to Greenland," scientists came together to dramatize what was at stake. Millions of marchers carrying signs and banners and linking arms made their case that a future that abandoned scientific truths would be a bleak one. It was a staggeringly compelling moment.

I was asked to be one of the speakers at the March for Science in Philadelphia, which, despite the cold and rain, produced twenty thousand marchers. The organizers had asked four local scientists to speak. Each of the other speakers stood at the podium and delivered rousing, evangelical speeches, stopping at key phrases to enjoy loud, enthusiastic applause. It was impressive. Here were scientists whose talks were usually punctuated with phrases like "Next slide, please," who were now delivering what could be characterized only as stump speeches. I was amazed that these other scientists could adapt to this situation so easily. And the crowds loved it.

My speech was a little tamer. When I walked up to the microphone and looked out at the throngs packing the amphitheater at Penn's Landing, the Delaware River behind me, I felt like I was at a rock concert—like something out of the movie *This Is Spinal Tap*. And like a scene in *Spinal Tap*, which was about a rock group in serious decline, I considered starting my talk with a loud, "Hello, Cleveland!" But I figured no one would get the joke, and there's nothing worse than beginning a talk with a bad joke (or ending a book with one).

Here's what I said:

> I was fortunate to participate with a team at the Children's Hospital of Philadelphia that created a vaccine that prevents a disease called rotavirus, a vaccine that has reduced the incidence of this disease in the U.S. by about 90 percent and is estimated to save hundreds of lives a day in the world.
>
> For twenty-five years, all of my funding came from the National Institutes of Health. I was funded by taxpayer dollars. In other words, the public paid my salary.
>
> If the public funds us, then we owe it to them to explain what we're doing, why we're doing it, and why it's important.
>
> Science is a privilege, not a right—and the taxpayer grants us that privilege.
>
> I'm encouraged to see so many people out today across the country and the world—an army of citizen scientists. Let's use this momentum to talk about science. To talk about how science has allowed us to live longer, better, healthier lives—how it has brought us out of the Age of Darkness and into the Age of Enlightenment.

We need to explain the importance of science to the media, to legislators, in church groups, PTA meetings, in elementary school classrooms, anyplace. No venue is too small. And by *we*, I mean anyone who is interested in science, including science teachers, science advocates, science enthusiasts, college students, high school students—anyone who loves science and can see what's at stake here.

We owe it to the public to explain ourselves. If not, I worry that in this age of anti-enlightenment, when science seems to be losing its place as a source of truth, we won't be able to do it for much longer.

So let's take the enthusiasm that we've created today and use it to excite everyone else in this country about the importance of science.

During the past twenty years, I have become friends with many parents and child advocates who work for organizations like the Autism Science Foundation, Every Child By Two, Families Fighting Flu, the Immunization Action Coalition, the Immunization Partnership, Meningitis Angels, the National Meningitis Association, Parents of Kids with Infectious Diseases, and Voices for Vaccines, among many others, all remarkably dedicated to the health and well being of children, and all remarkably generous with their time. It's heartening to know that so many wonderful people are out there. And I would never have met them had I not chosen to stand up for science.

ACKNOWLEDGMENTS

I would like to thank Patrick Fitzgerald for his patience and his guiding editorial hand. I would also like to thank Matt Carey, Larry Dubinski, Brian Fisher, Jeff Gerber, David Gorski, Eimear Kitt, Donald Mitchell, Charlotte Moser, Bonnie Offit, Emily Offit, Will Offit, Amy Pisani, Dorit Rubinstein Reiss, Anne Titterton, and Laura Vella for their helpful suggestions and careful reading of the manuscript.

APPENDIX:
BLOGS AND PODCASTS

Hundreds of blogs and podcasts provide reliable, accurate, and up-to-date information about science. Here I have listed those that I have found most useful in combating misinformation about vaccines, alternative medicine, climate change, autism, and a variety of other issues that dominate the news.

Discover Magazine Blogs: http://blogs.discovermagazine.com/
Harpocrates Speaks: www.harpocratesspeaks.com
Left Brain/Right Brain: https://leftbrainrightbrain.co.uk/
New Scientist: www.newscientist.com
Pharyngula: https://freethoughtblogs.com/pharyngula/
Popular Science: www.popsci.com/blog-network
Science Alert: www.sciencealert.com
Science-Based Medicine: https://sciencebasedmedicine.org
Science News: www.sciencenews.org/blogs
Scientific American Blogs: https://blogs.scientificamerican.com/
Skeptical Raptor: www.skepticalraptor.com/skepticalraptorblog.php
The Skeptics' Guide to the Universe: www.theskepticsguide.org
Virology Blog: www.virology.ws
Wired Science Blogs: www.wired.com/category/science-blogs/
You Are Not So Smart: https://youarenotsosmart.com/podcast

NOTES

Prologue: On Being Naïve

xii 1997 vaccine schedule: Centers for Disease Control and Prevention, "Notice to Readers: Recommended Childhood Immunization Schedule—United States, 1997," *Morbidity and Mortality Weekly Report* 46, no. 2 (1997) 46: 35–39.

1. What Science Is—and What It Isn't

1 Impact of scientific advances: Josh P. Bunker, Howard S. Frazier, and Frederick Mosteller, "Improving Health: Measuring Effects of Medical Care," *Milbank Quarterly* 72, no. 2 (1994): 225–58.

2 Opioid deaths: Centers for Disease Control and Prevention: Opioid deaths, https://www.cdc.gov/drugoverdose.

2 King quote: https:www.brainyquote.com/martin_luther_king_jr_102371.

4 Fibiger: Paul D. Stolley and Tamar Lasky, "Johannes Fibiger and His Nobel Prize for the Hypothesis that a Worm Causes Stomach Cancer," *Annals of Internal Medicine* 116, no. 9 (1992): 765–69.

4 Moniz: Jack El-Hai, *The Lobotomist: A Maverick Medical Genius and His Tragic Quest to Rid the World of Mental Illness* (Hoboken, NJ: John Wiley, 2005).

4 Keys: Olga Khazan, "When Trans Fats Were Healthy," *The Atlantic*, November 8, 2013, www.theatlantic.com/health/archive/2013/11/when-trans-fats-were-healthy/281274/.

5 MacMahon: Brian MacMahon, et al., "Coffee and Cancer of the Pancreas," *New England Journal of Medicine* 304, no. 11 (1981): 630–33.

5 Pons and Fleischmann: Gary Taubes, *Bad Science: The Short Life and Weird Times of Cold Fusion* (New York: Random House, 1993).

2. White Mice and Windowless Rooms

8   John Porter: Cornelia Dean, *Am I Making Myself Clear? A Scientist's Guide to Talking to the Public* (Cambridge, MA: Harvard University Press, 2009), 236.

9   Aaron Eckhart: David A. Kirby, *Lab Coats in Hollywood: Science, Scientists, and the Cinema* (Cambridge, MA: MIT Press, 2013), 72.

9   Margaret Mead: Christopher Frayling, *Mad, Bad and Dangerous? The Scientist and the Cinema* (London: Reaktion, 2005), 12.

10   David Chambers: Frayling, *Mad, Bad and Dangerous?*, 14.

10   Frayling on white lab coats: Frayling, *Mad, Bad and Dangerous?*, 18.

10   Milgram experiment: Stanley Milgram, *Obedience to Authority* (New York: Harper & Row, 1974).

11   Frayling on thick, black glasses: Frayling, *Mad, Bad, and Dangerous?*, 15.

11   Roszak on scientists: Frayling, *Mad, Bad, and Dangerous?*, 46.

11   Rarity of American scientists: Sidney Perkowitz, *Hollywood Science: Movies, Science, and the End of the World* (New York: Columbia University Press, 2007), 169.

13   Sagan and Salk: Annalee Newitz, "Why Was Carl Sagan Blackballed from the National Academy of Sciences?" *Gizmodo*, April 27, 2015, https://gizmodo.com/why-was-carl-sagan-blackballed-from-the-national-academ-1700524296; Paul A. Offit, *The Cutter Incident: How America's First Polio Vaccine Led to the Growing Vaccine Crisis* (New Haven, CT: Yale University Press, 2005).

15   Dan Burton: *Autism: Present Challenges, Future Needs—Why the Increased Rates? Hearing Before the Committee on Government Reform*, 106th Cong. (April 6, 2000) (statement of Rep. Dan Burton).

17   Thimerosal in vaccines: Paul A. Offit, *Autism's False Prophets: Bad Science, Risky Medicine, and the Search for a Cure* (New York: Columbia University Press, 2008), 81–129.

18   Neil deGrasse Tyson regarding *Titanic*: Maura Judkis, "'Titanic' Night Sky Adjusted After Neil deGrasse Tyson Criticized James Cameron," *Washington Post*, April 3, 2012.

22   Popularity of scientists: David J. Bennett and Richard C. Jennings, *Successful Science Communication: Telling It Like It Is* (Cambridge: Cambridge University Press, 2011), 226.

3. An Alibi for Ignorance

25   Bill Nye on Don Herbert: Bill Nye, "Teaching Science with a Big 'Poof,'" *Los Angeles Times*, June 15, 2007, http://articles.latimes.com/2007/jun/15/entertainment/et-wizard15.

25   Chondroitin sulfate and glucosamine: Daniel O. Clegg, et al., "Glucosamine, Chondroitin Sulfate, and the Two in Combination for Painful Knee Osteoarthritis," *New England Journal of Medicine* 354, no. 8 (2006): 795–808.

25 Concentrated garlic: Christopher D. Gardner, et al., "Effect of Raw Garlic vs. Commercial Garlic Supplements on Plasma Lipid Concentrations in Adults with Moderate Hypercholesterolemia: A Randomized Clinical Trial," *Archives of Internal Medicine* 167, no. 4 (2007): 346–53.

25 Saw palmetto: Stephen Bent, et al., "Saw Palmetto for Benign Prostatic Hyperplasia," *New England Journal of Medicine* 354, no. 6 (2006): 557–66; Michael J. Barry, et al., "Effect of Increasing Doses of Saw Palmetto Extract on Lower Urinary Tract Symptoms: A Randomized Trial," *Journal of the American Medical Association* 306, no. 12 (2011): 1344–51; David W. Freeman, "Saw Palmetto No Help for Enlarged Prostate, Study Says," *CBS News*, September 28, 2011, www.cbsnews.com/news/saw-palmetto -no-help-for-enlarged-prostate-study-says/.

26 Vitamin E: Eric A. Klein, et al., "Vitamin E and the Risk of Prostate Cancer: The Selenium and Vitamin E Cancer Prevention Trial (SELECT)," *Journal of the American Medical Association* 306, no. 14 (2011): 1549–56; Shirley S. Wang, "Is This the End of Popping Vitamins?" *Wall Street Journal*, October 25, 2011, www.wsj.com/articles/SB1000142405297020 4644504576650980601014152.

26 Ephedra: Shannon Brownlee, "Swallowing Ephedra," *Salon*, June 7, 2000, www.salon.com/2000/06/07/ephedra; Lorraine Heller, "Ephedra Recall Backed by Industry," *NutraIngredients*, April 13, 2008, www.nutrain gredients.com/Article/2008/04/14/Ephedra-recall-backed-by-industry.

26 Purity First vitamins: Delthia Ricks, "Recall of Purity First Vitamins Widens," *Newsday*, August 2, 2013, www.newsday.com/news/health/recall -of-purity-first-vitamins-widens-1.5821131.

26 OxyElite Pro: "OxyElite Pro Supplements Recalled," November 18, 2013, *Food and Drug Administration*, www.fda.gov/ForConsumers/Consumer Updates/ucm374742.htm.

26 Vitamin D intoxications: Cengiz Kara, et al., "Vitamin D Intoxication Due to an Erroneously Manufactured Dietary Supplement in Seven Children," *Pediatrics* 133, no. 1 (2014): e240–44.

26 Citrulline: "L-Citrulline by Medisca: Alert—Potentially Subpotent Product," *Food and Drug Administration*, February 14, 2014, www.fda. gov/Safety/MedWatch/Safetyinformation/SafetyAlertsforHumanMedical Products/ucm385978.htm.

27 Bo-Ying: "FDA Continues to Warn Consumers Not to Use Eu Yan Sang (Hong Kong) Ltd.'s 'Bo Ying Compound' " *Food and Drug Administration*, www.fda.gov/Drugs/DrugSafety/ucm416220.htm.

27 Acidophilus powder: Gary A. Emmett, "CDC Advisory: Infant Death Caused by Contaminated Probiotic Powder," *Philadelphia Inquirer*, December 4, 2014, www.philly.com/philly/blogs/healthy_kids/CDC -advisory-infant-death-caused-by-probiotic-powder.html; "FDA Investigates Presence of Mucormycosis-Causing Mold in Infant and Children's Probiotic Supplement," *Food and Drug Administration*, December 10, 2014; David Kroll, "Children's Probiotic Supplement Contaminated with

Disease-Causing Fungus," *Forbes*, November 18, 2014, www.forbes.com
/sites/davidkroll/2014/11/18/childrens-probiotic-supplement-contaminated
-with-disease-causing-fungus/#5ef562f9921f.

27　Sibutramine: "Dream Body Weight Loss Issues Voluntary Nationwide
Recall of Dream Body 450 mg, Dream Body Extreme Gold 800 mg,
Dream Body Advanced 400 mg Due to Undeclared Sibutramine," *Food
and Drug Administration*, July 1, 2016, www.fda.gov/Safety/Recalls
/ucm509727.htm.

28　Anti-vaccine celebrities: Anna Merlan, "Here's a Fairly Comprehensive
List of Anti-vaccination Celebrities," *Jezebel*, June 30, 2015, https://
jezebel.com/heres-a-fairly-comprehensive-list-of-anti-vaccination
-c-1714760128.

29　Reporter on cause and effect: Jeffrey Kluger, "RFK Jr. Joins the Anti-
vaccine Fringe," *Time*, July 21, 2014, http://time.com/3012797/vaccine
-rfk-jr-thimerosal/.

29　Keith Kloor article on RFK Jr: Keith Kloor, "Robert Kennedy Jr.'s Belief
in Autism–Vaccine Connection, and Its Political Peril," *Washington Post*,
July 18, 2014, www.washingtonpost.com/lifestyle/magazine/robert-kennedy
-jrs-belief-in-autism-vaccine-connection-and-its-political-peril/2014
/07/16/f21c01ee-f70b-11e3-a606-946fd632f9f1_story.html?utm_term
=.d04909615ff9.

30　Laura Helmuth on RFK Jr: Laura Helmuth, "Don't Feel Sorry for Robert F.
Kennedy Jr.," *Slate*, July 20, 2014, www.slate.com/articles/health_and
_science/science/2014/07/robert_f_kennedy_jr_profile_in_the_washington
_post_anti_vaccine_theory_and.html.

30　Jeffrey Kluger on RFK Jr: Kluger, "RFK Jr. Joins the Anti-vaccine Fringe."

31　Sherri Tenpenny on meningococcal B vaccine: "Student Guinea Pigs?"
NBC 10 Philadelphia, December 5, 2013, https://www.nbcphiladelphia
.com/news/health/Student-Guinea-Pigs__Philadelphia-234677351.html.

32　Sherri Tenpenny on Adam Lanza: "Sherri Tenpenny," *Rational Wiki*,
last modified December 28, 2016, https://rationalwiki.org/wiki/Sherri
_Tenpenny.

33　Journalism jail: David Kroll, "Dr. Paul Offit: 'Journalism Jail' for Faulty
Medical Reporting," *Forbes*, March 29, 2014, www.forbes.com/sites
/davidkroll/2014/03/29/dr-paul-offit-journalism-jail-for-false-equivalence
-medical-reporting/#5e111f591308.

35　Vaccine Injury Compensation Program: Paul A. Offit, *Deadly Choices:
How the Anti-vaccine Movement Threatens Us All* (New York: Basic
Books, 2011), 86–88.

36　Chapman University poll on false beliefs: Carrie Poppy, "Survey Shows
Americans Fear Ghosts, the Government, and Each Other," *Skeptical
Inquirer* 41, no. 1 (2017), www.csicop.org/si/show/survey_shows_americans
_fear_ghosts_the_government_and_each_other.

39　Bednarczyk study: Robert A. Bednarczyk, et al., "Sexual Activity–Related
Outcomes After Human Papillomavirus Vaccination of 11- to 12-Year-
Olds," *Pediatrics* 130, no. 5 (2012): 798–805.

40 Mike Pence and HPV vaccine: Associated Press, "Indiana Changes HPV Vaccine Notice After Criticism," *Washington Times*, November 6, 2015, www.washingtontimes.com/news/2015/nov/6/indiana-changes-hpv-vaccine-notice-after-criticism/; Jake Harper, "Public Health Experts Critical of Indiana's New HPV Vaccination Reminder," *WFYI Indianapolis*, November 10, 2015, www.wfyi.org/news/articles/public-health-experts-react-to-indianas-new-hpv-vaccination-reminder; Erica Hellerstein, "Mike Pence Put Ideology Before Science—And the People of Indiana Suffered," *ThinkProgress*, August 19, 2016, https://thinkprogress.org/mike-pence-public-health-406b5d08c7de/; Jake Harper, "As Indiana Governor, Mike Pence's Health Policy Has Been Contentious," *National Public Radio*, July 21, 2016, www.npr.org/sections/health-shots/2016/07/21/486771345/as-indiana-governor-mike-pence-s-health-policy-has-been-contentious; Abdul Hakim-Shabazz, "Health Community Petitions Pence on HPV Vaccine," *Indy Politics*, November 3, 2015, http://indypolitics.org/health-community-petitions-pence-on-hpv-vaccine/.

41 Yellow fever vaccine: Centers for Disease Control and Prevention, "Yellow Fever Vaccine: Recommendations of the Advisory Committee on Immunization Practices (ACIP)," *Morbidity and Mortality Weekly Report* 51, no. RR-17 (2002): 1–11.

45 Trump tweets about autism: Lindsay Dodgson, "Trump Has Suggested Vaccines Cause Autism—An Idea that Couldn't Be More Wrong," *Business Insider UK*, January 24, 2017, http://uk.businessinsider.com/trump-vaccines-autism-wrong-2017-1.

45 Trump meets with anti-vaccine activists: Casey Ross, "Andrew Wakefield Appearance at Trump Inaugural Ball Triggers Social Media Backlash," *STAT*, January 21, 2017, www.statnews.com/2017/01/21/andrew-wakefield-trump-inaugural-ball/.

45 Trump meets with RFK Jr: Abby Phillip, Lena H. Sun, and Lenny Bernstein, "Vaccine Skeptic Robert Kennedy Jr. Says Trump Asked Him to Lead Commission on 'Vaccine Safety,'" *Washington Post*, January 10, 2017, www.washingtonpost.com/politics/trump-to-meet-with-proponent-of-debunked-tie-between-vaccines-and-autism/2017/01/10/4a5d03c0-d752-11e6-9f9f-5cdb4b7f8dd7_story.html?utm_term=.8cc1c0af3f82.

47 GMOs: Wilhelm Klümper and Matin Qaim, "A Meta-analysis of the Impacts of Genetically Modified Crops," *PLOS One* 9, no. 11 (2014): e111629; Steven Novella, "No Health Risks from GMOs," *Skeptical Inquirer* 38, no. 4 (2014), www.csicop.org/si/show/no_health_risks_from_gmos.

47 Thomas Abinanti: Faye Flam, "Defying Science and Common Sense, New York Bill Would Ban GMOs in Vaccines," *Forbes*, February 26, 2015, www.forbes.com/sites/fayeflam/2015/02/26/defying-science-and-common-sense-new-york-bill-would-ban-gmos-in-vaccines/#40cdf6b6318e.

49 Gluten: Alan Levinovitz, *The Gluten Lie: And Other Myths About What You Eat* (New York: Regan Arts, 2015); Seamus O'Mahony, "A Postmodern Disease," *Dublin Review of Books*, February 1, 2017, www.drb.ie/essays/a-postmodern-disease.

49   Marc Vetri: Marc Vetri, "I'm Gluten Intolerant . . . Intolerant," *Huffington Post*, September 23, 2014, www.huffingtonpost.com/marc-vetri/im-gluten -intolerantintol_b_5614463.html.

4. Feeding the Beast

50   Declining media attention: David J. Bennett and Richard C. Jennings, *Successful Science Communication: Telling It Like It Is* (Cambridge: Cambridge University Press, 2011), 167; Cornelia Dean, *Am I Making Myself Clear? A Scientist's Guide to Talking to the Public* (Cambridge, MA: Harvard University Press, 2009), 38; Richard Hayes and Daniel Grossman, *A Scientist's Guide to Talking with the Media: Practical Advice from the Union of Concerned Scientists* (New Brunswick, NJ: Rutgers University Press, 2006), 32, 90.

52   Jenny McCarthy on *Oprah*: "Of Seizures and Celebrity: Evan's Grand-mother Speaks Up," *Just the Vax*, January 16, 2014, justthevax.blogspot .com/2014/01/of-seizures-and-celebrity-evans.html.

55   Oral polio vaccine didn't cause AIDS; Jon Cohen, "Forensic Epidemiology: Vaccine Theory of AIDS Origins Disputed at Royal Society," *Science* 289, no. 5486 (2000): 1850–51; B. Korber, et al., "Timing the Ancestor of the HIV-1 Pandemic Strains," *Science* 288, no. 5472 (2000): 1789–96; Michael Worobey, et al., "Origin of AIDS: Contaminated Polio Vaccine Theory Refuted," *Nature* 428, no. 6985 (2004): 820; Bruce Gellin, John F. Modlin, and Stanley A. Plotkin, "CHAT Oral Polio Vaccine Was Not the Source of Human Immunodeficiency Virus Type 1 Group M for Humans," *Clinical Infectious Diseases* 32, no. 7 (2001) 32: 1068–84; Michael Worobey, et al., "Direct Evidence of Extensive Diversity of HIV-1 in Kinshasa by 1960," *Nature* 455, no. 7213 (2008): 661–64.

56   *Good Day Philadelphia*: Tom Curtis, "The Origin of AIDS: A Startling New Theory Attempts to Answer the Question 'Was It an Act of God or an Act of Man?'" *Rolling Stone*, March 19, 1992; College of Physicians of Philadelphia, "Debunked: The Polio Vaccine and HIV Link," *The History of Vaccines*, http://historyofvaccines.org/content/articles/debunked-polio-vaccine-and-hiv-link; Paul M. Sharp and Beatrice H. Hahn, "Origins of HIV and the AIDS Pandemic," *Cold Spring Harbor Perspectives in Medicine* 1, no. 1 (2011): a006841; Jon Hilkevitz, "It's a Mystery Why Girl, 11, Has AIDS," *Chicago Tribune*, June 20, 1993, http://articles.chicagotribune.com /1993-06-20/news/9306200145_1_aids-virus-nurse-s-aide-vaccine; Jon Hilkevitz, "Case of Girl with AIDS Reopened," *Chicago Tribune*, July 15, 1993, http://articles.chicagotribune.com/1993-07-15/news/9307150165 _1_williams-case-williams-ads-bruce-williams; "Mystery of AIDS Girl Has a Twist," *Chicago Tribune*, July 18, 1993; Kenan Heise, "Whitney Williams, Young AIDS Victim," *Chicago Tribune*, May 17, 1997, http://articles .chicagotribune.com/1997-05-17/news/9705170153_1_young-aids -victim-identifiable-risk-sad.

57 *CBS This Morning*: "Don't Take Vitamins, Doctor Warns in New Book," *CBS News*, June 18, 2013, www.cbsnews.com/videos/dont-take-vitamins -doctor-warns-in-new-book/.

61 Inside Edition: "Could Megavitamins Be Bad for You?" *Inside Edition*, July 8, 2013, www.insideedition.com/consumer/6619-could-mega-vitamins -be-bad-for-you.

61 Gwyneth Paltrow and Goop: Kristine Phillips, "No, Gwyneth Paltrow, Women Should Not Put Jade Eggs in Their Vaginas, Gynecologist Says," *Washington Post*, January 22, 2017, www.washingtonpost.com/news/to -your-health/wp/2017/01/22/no-gwyneth-paltrow-women-should-not-put -jade-eggs-in-their-vaginas-gynecologist-says/; Meredith Engel, "Gwyneth Paltrow Wants Women to Steam Clean Their Vaginas—But Medical Experts Don't Recommend It," *New York Daily News*, January 29, 2015, www .nydailynews.com/life-style/health/gwyneth-paltrow-women-steam-vaginas -article-1.2096571; Katie Moisse, "Paltrow's 'Goop' Gets a Colon Cleanse," ABC News, January 6, 2012, http://abcnews.go.com/m/blogEntry?id =15306756.

68 John Kerridge on the media: Hayes and Grossman, *A Scientist's Guide*, 18.

69 Hepatitis B vaccine and SIDS: E. A. Mitchell, et al.,"Immunisation and the Sudden Infant Death Syndrome," *Archives of Disease in Childhood* 73, no. 6 (1995): 498–501; Manette T. Niu, Marcel E. Salive, and Susan S. Ellenberg, "Neonatal Deaths and Hepatitis B Vaccine: The Vaccine Adverse Event Reporting System, 1991–1998," *Archives of Pediatric and Adolescent Medicine* 153, no. 12 (1999): 1279–82; Eileen M. Eriksen, et al., "Lack of Association Between Hepatitis B Birth Immunization and Neonatal Death: A Population-Based Study from the Vaccine Safety DataLink Project," *Pediatric Infectious Disease Journal* 23, no. 7 (2004): 656–62.

69 Hepatitis B vaccine and multiple sclerosis: Alberto Ascherio, et al., "Hepatitis B Vaccination and the Risk of Multiple Sclerosis," *New England Journal of Medicine* 344, no. 5 (2001): 327–32; Christian Confavreux, et al., "Vaccinations and the Risk of Relapse in Multiple Sclerosis," *New England Journal of Medicine* 344, no. 5 (2001): 319–26.

70 HPV vaccine testing: J. T. Schiller, L. E. Markowitz, A. Hildesheim, and D. R. Lowry, "Human Papillomavirus Vaccines," in *Vaccines*, 7th edition, ed. S. A. Plotkin, W. A. Orenstein, P. A. Offit, and K. M. Edwards (Philadelphia: Elsevier, 2018).

71 Katie Couric and HPV vaccine: "The HPV Controversy," *Katie*, December 4, 2013, www.cbsnews.com.

71 Safety of HPV vaccine: C. Chao, et al., "Surveillance of Autoimmune Conditions Following Routine Use of Quadrivalent Human Papillomavirus Vaccine," *Journal of Internal Medicine* 271, no. 2 (2011): 193–203; Lisen Arnheim-Dahlström, et al., "Autoimmune, Neurological, and Venous Thromboembolic Adverse Events After Immunisation of Adolescent Girls

with Quadrivalent Human Papillomavirus Vaccine in Denmark and Sweden: Cohort Study," *British Medical Journal* 347 (2013): f5906; L. Grimaldi-Bensouda, et al., "Autoimmune Disorders and Quadrivalent Human Papillomavirus Vaccination of Young Female Subjects," *Journal of Internal Medicine* 275, no. 4 (2014): 398–408; Michelle Vichin, et al., "An Overview of Quadrivalent Human Papillomavirus Vaccine Safety: 2006 to 2015," *Pediatric Infectious Disease Journal* 34, no. 9 (2015): 983–91.

71   Pre-existing medical condition of patient on *Katie*: David Gorski, "Katie Couric on the HPV Vaccine: Antivaccine or Irresponsible Journalist? You Be the Judge!" *Respectful Insolence*, December 5, 2013, https:// respectfulinsolence.com/2013/12/05/katie-couric-on-the-hpv-vaccine -antivaccine-or-irresponsible-journalist-you-be-the-judge/.

72   Katie Couric's non-apology: Katie Couric, "Furthering the Conversation on the HPV Vaccine," *Huffington Post*, February 9, 2014, www.huffington post.com/katie-couric/vaccine-hpv-furthering-conversation_b_4418568 .html.

74   Heat treatment for AIDS: Hayes and Grossman, *A Scientist's Guide*, 34–36.

5. To Debate or Not to Debate

77   Studies of vaccinated versus unvaccinated children: Michael J. Smith and Charles R. Woods, "On-Time Vaccine Receipt in the First Year Does Not Adversely Affect Neuropsychological Outcomes," *Pediatrics* 125, no. 6 (2010): 1–8; Frank DeStefano, Cristofer S. Price, and Eric S. Weintraub, "Increasing Exposure to Antibody-Stimulating Proteins and Polysaccharides in Vaccines Is Not Associated with Risk of Autism," *Journal of Pediatrics* 163, no. 2 (2013): 561–67.

79   *Democracy Now!*: "Inside the Vaccine War: Measles Outbreak Rekindles Debate on Autism, Parental Choice and Public Health," *Democracy Now!*, February 5, 2015, www.democracynow.org/2015/2/5/inside_the_vaccine _war_measles_outbreak.

79   Richard Dawkins regarding the value of debate: Richard Dawkins, *A Devil's Chaplain: Reflections on Hope, Lies, Science, and Love* (Boston: Houghton Mifflin, 2003).

82   Bill Nye debates Ken Ham: David Freeman, "Bill Nye's Debate of Creationist Ken Ham Has Some Scientists Bothered," *Huffington Post*, February 4, 2014, www.huffingtonpost.ca/entry/bill-nye-debate-creationist -ken-ham-video_n_4714370; Pete Etchells, "Bill Nye v Ken Ham: Should Scientists Bother to Debate Creationism?" *The Guardian*, February 5, 2014, www.theguardian.com/science/head-quarters/2014/feb/05/bill-nye -vs-ken-ham-creationism-science-debate; Associated Press, "Bill Nye Defends Evolution in Kentucky Debate," *CBS News*, February 4, 2014, www .cbsnews.com/news/bill-nye-defends-evolution-in-kentucky-debate/;

Alan Boyle, "Bill Nye Wins Over the Science Crowd at Evolution Debate," *NBC News*, February 5, 2014, www.nbcnews.com/science/science-news /bill-nye-wins-over-science-crowd-evolution-debate-n22836; David Freeman, "Bill Nye Explains Why He Agreed to Debate Creationist Ken Ham," *Huffington Post*, April 14, 2014, www.huffingtonpost.ca/entry/bill-nye -debate-creationist-ken-ham_n_5147775.

86 Joe Schwarcz debates André Saine: "Debate About Homeopathy: Mere Placebo or Great Medicine?" YouTube video, 1:46:25, posted by André Saine, December 5, 2012, www.youtube.com/watch?v=T2uBBU4XT7Y.

87 Homeopathic ER: "That Mitchell and Webb Look: Homeopathic A&E," YouTube video, 2:33, posted by "gudbuytjane," July 2, 2009, www.youtube.com/watch?v=HMGIbOGu8q0.

88 Founding of the Institute for Historical Review: Michael Granberry, "Judge Awards $6.4 Million to O.C. Revisionist Group," *Los Angeles Times*, November 16, 1996, http://articles.latimes.com/1996-11-16/local /me-65105_1_judge-awards.

93 Michael Shermer debates Mark Weber: "A Holocaust Debate: Mark Weber vs Michael Shermer (full)," YouTube video, 1:55:53, posted by "Immortaltruthz," November 17, 2013, www.youtube.com/watch?v =4l8ZUVVB4z8.

6. Make 'Em Laugh

98 Franken quote: Al Franken, *Al Franken: Giant of the Senate* (New York: Twelve, 2017), 379.

99 Penn on "bullshit": Penn Gillette and Teller, "Talking to the Dead," *Penn & Teller: Bullshit!*, season 1, episode 1, directed by Star Price, aired January 24, 2003 (New York: Showtime).

100 Penn and Teller on vaccines: "Penn & Teller's Bullshit—Vaccinations," YouTube video, 1:30, posted by "Mkaidy," August 16, 2010, www.youtube .com/watch?v=lhk7-5eBCrs.

104 First appearance on *The Colbert Report*: "Dr. Paul Offit on the Stephen Colbert show," YouTube video, 5:28, posted by "IbrbSullivan," February 1, 2011, www.youtube.com/watch?v=KXntMFfrbjc.

105 Second appearance on *The Colbert Report*: "Preventable Diseases on the Rise," www.cc.com/video-clips/svsc0q/the-colbert-report-preventable -diseases-on-the rise.

109 Samantha Bee: "An Outbreak of Liberal Idiocy," http://www.cc.com /video-clips/g1lev1/the-daily-show-with-jon-stewart-an-outbreak-of -liberal-idiocy, May 2, 2014.

110 Zika virus in Florida: Centers for Disease Control and Prevention, "Zika Virus: 2016 Case Counts in the U.S.," *Centers for Disease Control and Prevention*, accessed February 26, 2017, www.cdc.gov/zika/reporting/2016 -case-counts.html.

110 Climate change: Brooks Hays, "Poll: 76 Percent of Americans Say Climate Change Is Happening," *UPI*, October 20, 2015, www.upi.com/Science _News/2015/10/20/Poll-76-percent-of-Americans-say-climate-change-is -happening/4111445374964/; Michael E. Mann, *The Hockey Stick and the Climate Wars: Dispatches from the Front Lines* (New York: Columbia University Press, 2012); Michael E. Mann and Tom Toles, *The Madhouse Effect: How Climate Change Denial Is Threatening Our Planet, Destroying Our Politics, and Driving Us Crazy* (New York: Columbia University Press, 2016).

112 John Oliver and climate change: "Climate Change on Last Week Tonight with John Oliver," https://www.youtube.com/watch?v=cjuGCJJUGsg, May 11, 2014.

114 John Oliver and dietary supplements: "Dr. Oz and Nutritional Supplements: Last Week Tonight with John Oliver (HBO)," YouTube video, 16:25, posted by "LastWeekTonight," June 22, 2014, www.youtube.com /watch?v=WA0wKeokWUU.

115 Jimmy Kimmel on vaccines: "A Message for the Anti-vaccine Movement," YouTube video, 4:54, posted by "Jimmy Kimmel Live," February 27, 2015, www.youtube.com/watch?v=QgpfNScEd3M.

116 Jimmy Kimmel's follow-up show on vaccines: "Jimmy Kimmel's Update on the Anti-vaccine Discussion," YouTube video, 6:49, posted by "Jimmy Kimmel Live," March 3, 2015, www.youtube.com/watch?v=i2mdwmp LYLY&t=11s.

7. Science Goes to the Movies

125 NASA consultants: David A. Kirby, *Lab Coats in Hollywood: Science, Scientists, and the Cinema* (Cambridge, MA: MIT Press, 2013), 51–53.

125 Other scientific consultants: Kirby, *Lab Coats in Hollywood*, 51–53.

126 Jack Horner: Kirby, *Lab Coats in Hollywood*, 234.

8. The Emperor's New Clothes

128 Tom Cruise takes on Brooke Shields: Tim Newcomb, "Tom Cruise at 50: Where Does the Newly Single Star Go from Here? The Brooke Shields Debacle," *Time*, June 28, 2012, http://entertainment.time.com/2012/07/03 /tom-cruise-at-50-where-does-the-controversial-star-go-from-here/slide /the-brook-shields-debacle/.

129 Matthew Carey: Mary Ann Roser, "Discredited Autism Guru Andrew Wakefield Takes Aim at the CDC," *Austin American-Statesman*, June 27, 2015, www.mystatesman.com/news/local/discredited-autism-guru-andrew -wakefield-takes-aim-cdc/7pBs9pYH4ssmaBXe6kQ1WM/.

129 Richard Horton comments on Andrew Wakefield: Richard Horton, *MMR Science and Fiction: Exploring a Vaccine Crisis* (London: Granta, 2004), 22.

130 Rosemary Kessick: Heather Mills, "MMR: The Story So Far," *Private Eye*, May 2002; Glenn Frankel, "Charismatic Doctor at Vortex of Vaccine Dispute," *Washington Post*, July 11, 2004, www.washingtonpost.com /wp-dyn/articles/A41450-2004Jul10.html; Grania Langdon-Down, "Law: A Shot in the Dark; The Complications from Vaccine Damage Seem to Multiply in the Courtroom," *The Independent*, November 27, 1996.

131 Articles about MMR, autism, and Wakefield: Paul A. Offit, *Autism's False Prophets: Bad Science, Risky Medicine, and the Search for a Cure* (New York: Columbia University Press, 2008), 18–36.

132 David Aaronovitch: David Aaronovitch, "Comment: A Travesty to Truth," *The Observer*, December 14, 2003.

133 Mady Hornig: Mady Hornig, et al., "Lack of Association Between Measles Vaccine Virus and Autism with Enteropathy: A Case-Control Study," *PLoS One* 3, no. 9 (2008): e3140.

134 Restrictive diets: Reviewed in S. Hurwitz, "The Gluten-Free, Casein-Free Diet and Autism: Limited Return on Family Investment," *Journal of Early Intervention* 35, no. 1 (2013): 3–19; "Thin Bones Seen in Boys with Autism and Autism Spectrum Disorder," press release, *National Institutes of Health*, January 29, 2008, www.nih.gov/news/health/jan2008/nichd-29.htm.

134 Studies exonerating MMR as a cause of autism: Heikki Peltola, et al., "No Evidence for Measles, Mumps, and Rubella Vaccine–Associated Inflammatory Bowel Disease or Autism in a 14-Year Prospective Study," *The Lancet* 351, no. 9112 (1998): 1327–28; Brent Taylor, et al., "Autism and Measles, Mumps, and Rubella Vaccine: No Epidemiological Evidence for a Causal Association," *The Lancet* 353, no. 9169 (1999): 2026–29; S. DeWilde, et al., "Do Children Who Become Autistic Consult More Often After MMR Vaccination?" *British Journal of General Practice* 51, no. 464 (2001): 226–27; Eric Fombonne and Suniti Chakrabarti, "No Evidence for a New Variant of Measles-Mumps-Rubella–Induced Autism," *Pediatrics* 108, no. 4 (2001): E58; Loring Dales, Sandra Jo Hammer, and Natalie J. Smith, "Time Trends in Autism and in MMR Immunization Coverage in California," *Journal of the American Medical Association* 285, no. 9 (2001): 1183–85; James A. Kaye, Maria del Mar Melero-Montes, and Hershel Jick, "Mumps, Measles, and Rubella Vaccine and the Incidence of Autism Recorded by General Practitioners: A Time Trend Analysis," *British Medical Journal* 322, no. 7284 (2001): 460–63; C. Paddy Farrington, Elizabeth Miller, and Brent Taylor, "MMR and Autism: Further Evidence Against a Causal Association," *Vaccine* 19, no. 27 (2001): 3632–35; N. Andrews, et al., "Recall Bias, MMR, and Autism," *Archives of Diseases of Children* 87, no. 6 (2002): 493–94; Annamari Mäkelä, J. Pekka Nuorti, and Heikki Peltola, "Neurologic Disorders After Measles–Mumps–Rubella Vaccination," *Pediatrics* 110, no. 5 (2002): 957–63; Brent Taylor, et al., "Measles, Mumps, and Rubella Vaccination and Bowel Problems or Developmental Regression in Children with

Autism: Population Study," *British Medical Journal* 324, no. 7334 (2002): 393–96; Kreesten Meldgaard Madsen, et al., "A Population-Based Study of Measles, Mumps, and Rubella Vaccination and Autism," *New England Journal of Medicine* 347, no. 19 (2002): 1477–82; Frank DeStefano, et al., "Age at First Measles–Mumps–Rubella Vaccination in Children with Autism and School-Matched Control Subjects: A Population-Based Study in Metropolitan Atlanta," *Pediatrics* 113, no. 2 (2004): 259–66; Liam Smeeth, et al., "MMR Vaccination and Pervasive Developmental Disorders: A Case-Control Study," *The Lancet* 364, no. 9438 (2004): 963–69; Tokio Uchiyama, Michiko Kurosawa, and Yutaka Inaba, "MMR-Vaccine and Regression in Autism Spectrum Disorders: Negative Results Presented from Japan," *Journal of Autism and Developmental Disorders* 37, no. 2 (2007): 210–17; Mady Hornig, et al., "Lack of Association Between Measles Virus Vaccine and Autism with Enteropathy: A Case-Control Study," *PLoS One* 3, no. 9 (2008): e3140; Dorota Mrozek-Budzyn, Agnieszka Kieltyka, and Renata Majewska, "Lack of Association Between Measles-Mumps-Rubella Vaccination and Autism in Children: A Case-Control Study," *Pediatric Infectious Disease Journal* 29, no. 5 (2010): 397–400; Yota Uno, et al., "The Combined Measles, Mumps, and Rubella Vaccines and the Total Number of Vaccines Are Not Associated with Development of Autism Spectrum Disorder: The First Case-Control Study in Asia," *Vaccine* 30, no. 28 (2012): 4292–98; Luke E. Taylor, Amy L. Swerdfeger, and Guy D. Eslick, "Vaccines Are Not Associated with Autism: An Evidence-Based Meta-Analysis of Case-Control and Cohort Studies," *Vaccine* 32, no. 29 (2014): 3623–29; Anjali Jain, et al., "Autism Occurrence by MMR Vaccine Status Among U.S. Children with Older Siblings with and Without Autism," *Journal of the American Medical Association* 313, no. 15 (2015): 1534–40.

134 Wakefield on resignation: Lorraine Fraser, "Anti-MMR Doctor Is Forced Out," *Sunday Telegraph*, December 2, 2001, www.telegraph.co.uk/news /uknews/1364080/Anti-MMR-doctor-is-forced-out.html.

135 Wakefield on "bigger issues": Fraser, "Anti-MMR Doctor Is Forced Out."

136 Thoughtful House: A. Ahuja, "MMR Maverick," *The Times* (London), June 13, 2006; Michael Fitzpatrick, *MMR and Autism: What Parents Need to Know* (London: Routledge, 2004), 163; Horton, *MMR Science and Fiction*, 131.

137 Brian Deer and the GMC: John F. Burns, "British Medical Council Bars Doctor Who Linked Vaccine with Autism," *New York Times*, May 24, 2010, www.nytimes.com/2010/05/25/health/policy/25autism.html; Alice Park, "Doctor Behind Vaccine–Autism Link Loses License," *Time*, May 24, 2010, http://healthland.time.com/2010/05/24/doctor-behind-vaccine-autism -link-loses-license/; "Autism Study Doctor Barred for 'Serious Misconduct,'" *CNN*, May 24, 2010, www.cnn.com/2010/HEALTH/05/24/autism.vaccine .doctor.banned/index.html.

137 Wakefield on "not going away": Michael Inbar, "Controversial Autism Doc: 'I'm Not Going Away,'" May 24, 2010, *Today*, www.today.com /health/controversial-autism-doc-im-not-going-away-1C9400946.

137 Rebecca Estep: Madison Park, "Medical Journal Retracts Study Linking Autism to Vaccine," *CNN*, February 2, 2010, www.cnn.com/2010 /HEALTH/02/02/lancet.retraction.autism/index.html.

137 *Lancet* retracts Wakefield's paper: Gardiner Harris, "Journal Retracts 1998 Paper Linking Autism to Vaccines," *New York Times*, February 2, 2010, www.nytimes.com/2010/02/03/health/research/03lancet.html; Sarah Boseley, "*Lancet* Retracts 'Utterly False' MMR Paper," *The Guardian*, February 2, 2010, www.theguardian.com/society/2010/feb/02/lancet-retracts -mmr-paper; Park, "Medical Journal Retracts Study Linking Autism to Vaccine"; Laura Eggerston, "*Lancet* Retracts 12-Year-Old Article Linking Autism to MMR Vaccines," *Canadian Medical Association Journal* 182, no. 4 (2010): E199–200; Fiona Godlee, Jane Smith, and Harvey Marcovitch. "Wakefield's Article Linking MMR Vaccine and Autism Was Fraudulent," *British Medical Journal* 342 (2011): c7452.

137 Horton on Wakefield's paper: Boseley, "*Lancet* Retracts 'Utterly False' MMR Paper."

138 Godlee on Wakefield: Godlee, Smith, and Marcovitch, "Wakefield's Article Linking MMR Vaccine and Autism Was Fraudulent."

139 Anderson Cooper interview: "Retracted Autism Study an 'Elaborate Fraud,' British Journal Finds," *Anderson Cooper 360°*, January 5, 2011, http://ac360.blogs.cnn.com/2011/01/05/retracted-autism-study-an-elaborate -fraud-british-journal-finds/?hpt=ac_mid.

140 The ConspiraSea Cruise: April Glaser, "A Skeptic Infiltrates a Cruise for Conspiracy Theorists," *Wired*, February 9, 2016, www.wired.com/2016 /02/conspira-sea-cruise-know-truth/; Anna Merlan, "Sail (Far) Away. At Sea with America's Largest Floating Gathering of Conspiracy Theorists," *Jezebel*, February 25, 2016, https://jezebel.com/sail-far-away-at-sea-with -americas-largest-floating-1760900554.

141 Isabella Thomas: Horton, *MMR Science and Fiction*, 9.

142 Wakefield on colleagues: Merlan, "Sail (Far) Away."

142 Wakefield on GMC hearing: Harrison, J., "Wrong About Vaccine Safety: A Review of Andrew Wakefield's 'Callous Disregard,'" *Open Vaccine Journal* 6 (2013): 9–25.

143 *Callous Disregard*: Andrew J. Wakefield, *Callous Disregard: Autism and Vaccines—The Truth Behind a Tragedy* (New York: Skyhorse, 2010).

144 Wakefield and "Through a Glass Darkly": Andrew J. Wakefield and Scott M. Montgomery, "Measles, Mumps, Rubella Vaccine: Through a Glass, Darkly," *Adverse Drug Reactions and Toxicological Reviews* 19, no. 4 (2000): 265–83.

144 Wakefield at Trinity Christian College: Ken Reibel, personal communication by email.

144 Wakefield on "inheriting the earth": Andrew J. Wakefield, "Chiropractors and Vaccines: Where Should We Stand?" Presentation given at California Jam, a chiropractic education and natural health seminar, Costa Mesa, CA, March 18–20, 2016.

144 Wakefield on dying for children with autism: Andrew J. Wakefield, *Autism Media Channel*, www.autismmediachannel.com/.

144 Handley regarding Wakefield: Susan Dominus, "The Crash and Burn of an Autism Guru," *New York Times Magazine*, April 20, 2011, www.nytimes .com/2011/04/24/magazine/mag-24Autism-t.html.

145 Numbers of unvaccinated children in the United States: Centers for Disease Control and Prevention, "National, State, and Selected Local Area Vaccination Coverage Among Children Aged 19–35 Months—United States, 2013," *Morbidity and Mortality Weekly Report* 63, no. 34 (2014): 741–48.

145 2014 measles epidemic: Lenny Bernstein, "U.S. Measles Outbreak Sets Record for Post-elimination Era," *Washington Post*, May 29, 2014, www.washingtonpost.com/news/to-your-health/wp/2014/05/29 /u-s-measles-outbreak-sets-record-for-post-elimination-era/?utm_term =.556aacee3807.

145 2015 measles epidemic: Centers for Disease Control and Prevention, "Measles Outbreak—California, December 2014–February 2015," *Morbidity and Mortality Weekly Report* 64, no. 6 (2015): 153–54.

145 2016 mumps epidemic: "Mumps Cases and Outbreaks," *Centers for Disease Control and Prevention*, accessed February 26, 2017, www.cdc .gov/mumps/outbreaks.html.

145 2017 measles outbreak among Somalis in Minnesota: Paul A. Offit, "Did Anti-vaxxers Spark a Measles Outbreak in an Immigrant Community?" *The Daily Beast*, May 13, 2017, www.thedailybeast.com/did-anti-vaxxers -spark-a-measles-outbreak-in-an-immigrant-community.

145 Wakefield on responsibility: Lena H. Sun, "Anti-vaccine Activists Spark a State's Worst Measles Outbreak in Decades," *Washington Post*, May 5, 2017, www.washingtonpost.com/national/health-science/anti-vaccine-activists-spark-a-states-worst-measles-outbreak-in-decades/2017/05/04 /a1fac952-2f39-11e7-9dec-764dc781686f_story.html?utm_term =.25173596658a; Lena H. Sun, "Despite Measles Outbreaks, Anti-vaccine Activists in Minnesota Refuse to Back Down," *Washington Post*, August 21, 2017, www.washingtonpost.com/national/health-science/despite-measles -outbreak-anti-vaccine-activists-in-minnesota-refuse-to-back-down/2017 /08/21/886cca3e-820a-11e7-ab27-1a21a8e006ab_story.html?utm_term =.4ae888ad3752.

9. Judgment Day

146 Burton hearing: *Autism: Present Challenges, Future Needs—Why the Increased Rates? Hearing Before the Committee on Government Reform*, 106th Cong. (April 6, 2000).

149 Hornig study: Mady Hornig, et al., "Lack of Association Between Measles Vaccine Virus and Autism with Enteropathy: A Case-Control Study," *PLoS One* 3, no. 9 (2008): e3140.

152 Buck Offit: Sidney Offit, *Memoir of the Bookie's Son* (New York: St. Martin's, 1995).

10. The Nuclear Option

166 Amy Wallace and *Wired*: Amy Wallace, "An Epidemic of Fear," *Wired*, November 2009.

168 Barbara Loe Fisher lawsuit: Fisher v. Offit, *Digital Media Law Project*, May 6, 2010, www.dmlp.org/threats/fisher-v-offit.

169 Richard Barr lawsuit: Clare Dyer, "Libel Threat Delays UK Publication of Book on Campaigns Against Vaccines," *British Medical Journal* 342 (2011): 1159.

170 Popehat: Ken White, "So You've Been Threatened with a Defamation Suit," *Popehat*, September 26, 2013, www.popehat.com/2013/09/26/so -youve-been-threatened-with-a-defamation-suit/.

171 Rally at NYU Langone conference: "NYU Holds a Conference on Resistance to Vaccines: Antivaxers Lose It," Respectful Insolence, https:// respectfulinsolence.com/2016/11/22/nyu-holds-a-conference, November 22, 2016.

171 Polly Tommey: Tom Whipple, "Showing Anti-vaccine Film 'Is a Health Risk,'" *The Times*, April 28, 2017, www.thetimes.co.uk/article/2c948a68 -2b7e-11e7-9d2e-96f2194e0ac4.

173 Joshua Coleman: Ben Deci, "Roseville Anti-vaccination Campaigner Charged with Willful Cruelty to Children," *Fox40*, October 30, 2015, http://fox40.com/2015/10/30/roseville-anti-vaccination-campaigner-charged -with-willful-cruelty-to-children/.

11. Pharma Shill

178 RFK Jr. vilifications: Keith Kloor, "Is Robert F. Kennedy Jr. Anti-science?" *Discover*, June 1, 2013, http://blogs.discovermagazine.com/collideascape /2013/06/01/is-robert-f-kennedy-jr-anti-science/#.Wk09v9-nE2w.

178 Jim Carrey vilifications: Jim Carrey, "The Judgment on Vaccines Is In???" *Huffington Post*, November 17, 2011, www.huffingtonpost.com/jim-carrey /the-judgment-on-vaccines_b_189777.html.

178 Jenny McCarthy: Jenny McCarthy, Twitter post, February 27, 2010, 2:00 p.m., https://twitter.com/jennymccarthy/status/9741176086.

178 Chelsea Handler: Chelsea Handler, Twitter post, April 1, 2010, 2:38 p.m., https://twitter.com/chelseahandler/status/11438807878.

186 Rotavirus vaccine Phase III trial: Timo Vesikari, et al., "Safety and Efficacy of Pentavalent Human–Bovine (WC3) Reassortant Rotavirus Vaccine," *New England Journal of Medicine* 354, no. 1 (2006): 23–33.

187 Earlier rotavirus vaccine and intussusception: Trudy V. Murphy, et al., "Intussusception Among Infants Given an Oral Rotavirus Vaccine," *New England Journal of Medicine* 344, no. 8 (2001): 564–72.

190 MMR and autism: Heikki Peltola, et al., "No Evidence for Measles, Mumps, and Rubella Vaccine–Associated Inflammatory Bowel Disease or Autism in a 14-Year Prospective Study," *The Lancet* 351, no. 9112 (1998): 1327–28; Brent Taylor, et al., "Autism and Measles, Mumps, and Rubella Vaccine: No Epidemiological Evidence for a Causal Association," *The Lancet* 353, no. 9169 (1999): 2026–29; S. DeWilde, et al., "Do Children Who Become Autistic Consult More Often After MMR Vaccination?" *British Journal of General Practice* 51, no. 464 (2001): 226–27; Eric Fombonne and Suniti Chakrabarti, "No Evidence for a New Variant of Measles-Mumps-Rubella–Induced Autism," *Pediatrics* 108, no. 4 (2001): E58; Loring Dales, Sandra Jo Hammer, and Natalie J. Smith, "Time Trends in Autism and in MMR Immunization Coverage in California," *Journal of the American Medical Association* 285, no. 9 (2001): 1183–85; James A. Kaye, Maria del Mar Melero-Montes, and Hershel Jick, "Mumps, Measles, and Rubella Vaccine and the Incidence of Autism Recorded by General Practitioners: A Time Trend Analysis," *British Medical Journal* 322, no. 7284 (2001): 460–63; C. Paddy Farrington, Elizabeth Miller, and Brent Taylor, "MMR and Autism: Further Evidence Against a Causal Association," *Vaccine* 19, no. 27 (2001): 3632–35; N. Andrews, et al., "Recall Bias, MMR, and Autism," *Archives of Diseases of Children* 87, no. 6 (2002): 493–94; Annamari Mäkelä, J. Pekka Nuorti, and Heikki Peltola, "Neurologic Disorders After Measles–Mumps–Rubella Vaccination," *Pediatrics* 110, no. 5 (2002): 957–63; Brent Taylor, et al., "Measles, Mumps, and Rubella Vaccination and Bowel Problems or Developmental Regression in Children with Autism: Population Study," *British Medical Journal* 324, no. 7334 (2002): 393–96; Kreesten Meldgaard Madsen, et al., "A Population-Based Study of Measles, Mumps, and Rubella Vaccination and Autism," *New England Journal of Medicine* 347, no. 19 (2002): 1477–82; Frank DeStefano, et al., "Age at First Measles–Mumps–Rubella Vaccination in Children with Autism and School-Matched Control Subjects: A Population-Based Study in Metropolitan Atlanta," *Pediatrics* 113, no. 2 (2004): 259–66; Liam Smeeth, et al., "MMR Vaccination and Pervasive Developmental Disorders: A Case-Control Study," *The Lancet* 364, no. 9438 (2004): 963–69; Tokio Uchiyama, Michiko Kurosawa, and Yutaka Inaba, "MMR-Vaccine and Regression in Autism Spectrum Disorders: Negative Results Presented from Japan," *Journal of Autism and Developmental Disorders* 37, no. 2 (2007): 210–17; Mady Hornig, et al., "Lack of Association Between Measles Virus Vaccine and Autism with Enteropathy: A Case-Control Study," *PLoS One* 3, no. 9 (2008): e3140; Dorota Mrozek-Budzyn, Agnieszka Kieltyka, and Renata Majewska, "Lack of Association Between Measles-Mumps-Rubella Vaccination and Autism in Children: A Case-Control Study," *Pediatric Infectious Disease*

Journal 29, no. 5 (2010): 397–400; Yota Uno, et al., "The Combined Measles, Mumps, and Rubella Vaccines and the Total Number of Vaccines Are Not Associated with Development of Autism Spectrum Disorder: The First Case-Control Study in Asia," *Vaccine* 30, no. 28 (2012): 4292–98; Luke E. Taylor, Amy L. Swerdfeger, and Guy D. Eslick, "Vaccines Are Not Associated with Autism: An Evidence-Based Meta-Analysis of Case-Control and Cohort Studies," *Vaccine* 32, no. 29 (2014): 3623–29; Anjali Jain, et al., "Autism Occurrence by MMR Vaccine Status Among U.S. Children with Older Siblings with and Without Autism," *Journal of the American Medical Association* 313, no. 15 (2015): 1534–40.

190 Thimerosal and autism: Paul Stehr-Green, et al., "Autism and Thimerosal-Containing Vaccines: Lack of Consistent Evidence for an Association," *American Journal of Preventive Medicine* 25, no. 2 (2003): 101–06; Kreesten M. Madsen, et al., "Thimerosal and the Occurrence of Autism: Negative Ecological Evidence from Danish Population-Based Data," *Pediatrics* 112, no. 3, part 1 (2003): 604–06; Anders Hviid, et al., "Association Between Thimerosal-Containing Vaccines and Autism," *Journal of the American Medical Association* 290, no. 13 (2003): 1763 66; Jon Heron, Jean Golding, and the ALSPAC Study Team, "Thimerosal Exposure in Infants and Developmental Disorders: A Prospective Cohort Study in the United Kingdom Does Not Support a Causal Association," *Pediatrics* 114, no. 3 (2004): 577–83; Nick Andrews, et al., "Thimerosal Exposure in Infants and Developmental Disorders: A Retrospective Cohort Study in the United Kingdom Does Not Support a Causal Association," *Pediatrics* 114, no. 3 (2004): 584–91; Eric Fombonne, et al., "Pervasive Developmental Disorders in Montreal, Quebec, Canada: Prevalence and Links with Immunizations," *Pediatrics* 118, no. 1 (2006): 139–50; William W. Thompson, et al., "Early Thimerosal Exposure and Neuropsychological Outcomes at 7 to 10 Years," *New England Journal of Medicine* 357, no. 13 (2007): 1281–92; Robert Schechter and Judith K. Grether, "Continuing Increases in Autism Reported to California's Development Services System: Mercury in Retrograde," *Archives of General Psychiatry* 65, no. 1 (2008): 19–24.

190 Too many vaccines and autism: Michael J. Smith and Charles R. Woods, "On-Time Vaccine Receipt in the First Year Does Not Adversely Affect Neuropsychological Outcomes," *Pediatrics* 125, no. 6 (2010): 1134–41; Frank DeStefano, Cristofer S. Price, and Eric S. Weintraub, "Increasing Exposure to Antibody-Stimulating Proteins and Polysaccharides in Vaccines Is Not Associated with Risk of Autism," *Journal of Pediatrics* 163, no. 2 (2013): 561–67.

194 *Deadly Immunity*: Robert F. Kennedy Jr., "Deadly Immunity," *Rolling Stone*, June 30–July 14, 2005.

194 Salon.com: Kerry Lauerman, "Correcting Our Record," *Salon*, January 16, 2011, www.salon.com/2011/01/16/dangerous_immunity/.

195 Sharyl Attkisson report: Sharyl Attkisson, "How Independent Are Vaccine Defenders?" *CBS News*, July 25, 2008, www.cbsnews.com/news/how -independent-are-vaccine-defenders/.

198 Matt Lauer and *Dateline NBC*: "Dose of Controversy," *Dateline NBC*, August 30, 2009.

12. A Ray of Hope

201 *Vaxxed* promo: David Gorski, "What's Going On Here? Andrew Wakefield's Anti-vaccine Propaganda Film to Be Screened at the Tribeca Film Festival," *Respectful Insolence*, March 22, 2016, https://respectful insolence.com/2016/03/22/wtf-andrew-wakefields-antivaccine-documentary -to-be-screened-at-the-tribeca-film-festival/; Laura June, "Why Is an Anti-vaccine Documentary by a Proven Quack Being Taken Seriously?" *The Cut*, March 22, 2016, www.thecut.com/2016/03/tribeca-film-festival -anti-vaxx.html.

201 De Niro premiers *Vaxxed*: Jeremy Gerard, "Robert De Niro Defends 'Personal' Decision to Screen Anti-vaccination Docu at Tribeca Film Fest as Protests Mount," *Deadline Hollywood*, March 25, 2016.

202 Michael Specter: Steven Zeitchik, "Tribeca to Screen Movie by Controversial Anti-vaccine Activist Andrew Wakefield," *Los Angeles Times*, March 22, 2016, www.latimes.com/entertainment/movies/moviesnow /la-et-mn-anti-vaccine-andrew-wakefield-movie-tribeca-20160322-story .html.

202 Michael Hiltzik: Michael Hiltzik, "How Robert De Niro's Tribeca Film Festival Sold Out to Anti-vaccine Crackpots," *Los Angeles Times*, March 25, 2016, www.latimes.com/business/hiltzik/la-fi-hiltzik-tribeca -vaccine-20160323-snap-htmlstory.html.

202 Alexandra Sowa: Alexandra Sowa, "Robert De Niro Should Take Heat for Giving Tribeca Film Festival Platform to Anti-vaccination Documentary by Snake-Oil Peddler Andrew Wakefield," *New York Daily News*, March 26, 2016, www.nydailynews.com/opinion/robert-de-niro-giving -tribeca-platform-vaccination-fraud-article-1.2578692.

203 Colin McRoberts on *Vaxxed*: Colin McRoberts, "Vaxxed: From Cover-Up to Catastrophe to Cancellation to Insignificance," *Violent Metaphors*, March 31, 2016, http://violentmetaphors.com/2016/03/31/vaxxed-from -coverup-to-catastrophe-to-cancellation-to-insignificance.

203 "DPT Vaccine Roulette": "DPT: Vaccine Roulette," WRC-TV, Washington, DC, April 19, 1982.

203 Studies exonerating pertussis vaccine as a cause of brain damage: Christopher L. Cody, et al., "Nature and Rates of Adverse Reactions Associated with DTP and DT Immunizations in Infants and Children," *Pediatrics* 68, no. 5 (1981): 650–60; Samuel Bedson, et al., "Vaccination Against Whooping Cough: Relation Between Protection in Children and Results of Laboratory Tests. A Report to the Whooping Cough Immunization

Committee of the Medical Research Council and to the Medical Officers of Health for Cardiff, Leeds, Leyton, Manchester, Middlesex, Oxford, Poole, Tottenham, Walthamstow, and Wembley," *British Medical Journal* 2, no. 4990 (1956) 2: 454–62; Bo Hellström, "Electroencephalographic Studies in Triple-Immunized Infants," *British Medical Journal* 2, no. 5312 (1962): 1089–91; T. M. Pollack and Jean Morris, "A 7-Year Survey of Disorders Attributed to Vaccination in North West Thames Region," *The Lancet* 1, no. 8327 (1983): 753–57; J. A. N. Corsellis, I. Janota, and A. K. Marshall, "Immunization Against Whooping Cough: A Neuropathological Review," *Neuropathology and Applied Neurobiology* 9, no. 4 (1983): 261–70; W. Donald Shields, et al., "Relationship of Pertussis Immunization to the Onset of Neurologic Disorders: A Retrospective Epidemiologic Study," *Journal of Pediatrics* 113, no. 5 (1988): 801–05; Alexander M. Walker, et al., "Neurologic Events Following Diphtheria–Tetanus–Pertussis Immunization," *Pediatrics* 81, no. 3 (1988): 345–49; Marie R. Griffin, et al., "Risk of Seizures and Encephalopathy After Immunization with the Diphtheria–Tetanus–Pertussis Vaccine," *Journal of the American Medical Association* 263, no. 12 (1990): 1641–45; James D. Cherry, " 'Pertussis Vaccine Encephalopathy': It Is Time to Recognize It as the Myth that It Is," *Journal of the American Medical Association* 263, no. 12 (1990): 1679–80; Gerald S. Golden, "Pertussis Vaccine and Injury to the Brain," *Journal of Pediatrics* 116, no. 6 (1990): 854–61; Institute of Medicine, *Adverse Effects of Pertussis and Rubella Vaccines: A Report of the Committee to Review the Adverse Consequences of Pertussis and Rubella Vaccines* (Washington, DC: National Academies Press, 1991); Ad Hoc Committee for the Child Neurology Consensus Statement on Pertussis Immunization and the Central Nervous System, "Pertussis Immunization and the Central Nervous System," *Annals of Neurology* 29, no. 4 (1991): 458–60; James L. Gale, et al., "Risk of Serious Acute Neurological Illness After Immunization With Diphtheria–Tetanus–Pertussis Vaccine: A Population-Based Case-Control Study," *Journal of the American Medical Association* 271, no. 1 (1994): 37–41; William E. Barlow, et al., "The Risk of Seizures After Receipt of Whole-Cell Pertussis or Measles, Mumps, and Rubella Vaccine," *New England Journal of Medicine* 345, no. 9 (2001): 656–61; David Miller, et al., "Pertussis Immunisation and Serious Acute Neurological Illness in Children," *British Medical Journal* 282, no. 6276 (1981): 1595–99.

203 Fisher and Hib vaccine: *ABC World News Tonight with Peter Jennings*, February 16, 1998.

204 Fisher and ABC's *20/20*: *20/20*, January 22, 1999.

204 Fisher and pneumococcal vaccine: "New Pneumococcal Vaccines for Children," *World News Tonight with Peter Jennings*, September 29, 1998.

204 J. B. Handley: Paul A. Offit, *Deadly Choices: How the Anti-vaccine Movement Threatens Us All* (New York: Basic Books, 2011), 154–62.

205 *Hollywood Reporter* review of *Vaxxed*: Paul Offit, "Anti-Vaccine Doc 'Vaxxed': A Doctor's Film Review," *Hollywood Reporter*, April 11, 2016, www.hollywoodreporter.com/news/anti-vaccine-doc-vaxxed-a-882651.

207 Ken Reibel: Mary Ann Roser, "Discredited Autism Guru Andrew Wakefield Takes Aim at CDC," *Austin American-Statesman*, June 27, 2015, www.mystatesman.com/news/local/discredited-autism-guru-andrew -wakefield-takes-aim-cdc/7pBs9pYH4ssmaBXe6kQ1WM/.

207 Matthew Carey: Roser, "Discredited Autism Guru."

207 Parents' views on the relationship between vaccines and autism: Marina Sarris, "Bridging the Divide Between Parents and Scientists," *Interactive Autism Network*, March 12, 2015, https://iancommunity.org/ssc/bridging -divide-between-parents-scientists.

208 McRoberts on Wakefield: McRoberts, "Vaxxed: From Cover-Up to Catastrophe."

Epilogue: The End of the Tour

209 March for Science: Seth Borenstein, "Brandenburg Gate to the Washington Monument: Worldwide Marches Push for Science Over Politics" *Associated Press*, April 22, 2017.

SELECTED BIBLIOGRAPHY

Baron, Nancy. *Escape from the Ivory Tower: A Guide to Making Your Science Matter*. Washington, DC: Island, 2010.

Bennett, David J., and Richard C. Jennings. *Successful Science Communication: Telling It Like It Is*. Cambridge: Cambridge University Press, 2011.

Berezow, Alex B., and Hank Campbell. *Science Left Behind: Feel-Good Fallacies and the Rise of the Anti-Scientific Left*. New York: Public Affairs, 2012.

Bernstein, Peter L. *Against the Gods: The Remarkable Story of Risk*. New York: Wiley, 1996.

Dean, Cornelia. *Am I Making Myself Clear? A Scientist's Guide to Talking to the Public*. Cambridge, MA: Harvard University Press, 2009.

Frayling, Christopher. *Mad, Bad and Dangerous? The Scientist and the Cinema*. London: Reaktion, 2005.

Hayes, Richard, and Daniel Grossman. *A Scientist's Guide to Talking with the Media: Practical Advice from the Union of Concerned Scientists*. New Brunswick, NJ: Rutgers University Press, 2006.

Horton, Richard. *MMR Science and Fiction: Exploring the Vaccine Crisis*. London: Granta, 2004.

Kirby, David A. *Lab Coats in Hollywood: Science, Scientists, and the Cinema*. Cambridge, MA: MIT Press, 2013.

Luna, Rafael L. *The Art of Scientific Storytelling: Transform Your Research Manuscript with a Step-By-Step Formula*. Middletown, DE: Amado International, 2013.

Mann, Michael E. *The Hockey Stick and the Climate Wars: Dispatches from the Front Lines*. New York: Columbia University Press, 2012.

Mann, Michael E., and Tom Toles. *The Madhouse Effect: How Climate Change Denial Is Threatening Our Planet, Destroying Our Politics, and Driving Us Crazy*. New York: Columbia University Press, 2016.

Mooney, Chris. *The Republican War on Science*. New York: Basic Books, 2005.

Offit, Paul A. *Autism's False Prophets: Bad Science, Risky Medicine, and the Search for a Cure*. New York: Columbia University Press, 2008.

——. *Bad Faith: When Religious Belief Undermines Modern Medicine*. New York: Basic Books, 2015.

——. *Deadly Choices: How the Anti-vaccine Movement Threatens Us All*. New York: Basic Books, 2011.

——. *Do You Believe in Magic? The Sense and Nonsense of Alternative Medicine*. New York: HarperCollins, 2013.

——. *Pandora's Lab: Seven Stories of Science Gone Wrong*. Washington DC: National Geographic, 2017.

Olson, Randy. *Don't Be Such a Scientist: Talking Substance in an Age of Style*. Washington, DC: Island, 2009.

——. *Houston, We Have a Narrative*. Chicago: University of Chicago Press, 2015.

Olson, Randy, Dorie Barton, and Brian Palermo. *Connection: Hollywood Storytelling Meets Critical Thinking*. Los Angeles: Prairie Starfish, 2013.

Perkowitz, Sidney. *Hollywood Science: Movies, Science, and the End of the World*. New York: Columbia University Press, 2007.

Russell, Nicholas. *Communicating Science: Professional, Popular, Literary*. Cambridge: Cambridge University Press, 2010.

Shermer, Michael. *Why People Believe Weird Things: Pseudoscience, Superstition, and Other Confusions of Our Time*. New York: W. H. Freeman, 1997.

Shermer, Michael, and Alex Grobman. *Denying History: Who Says the Holocaust Never Happened and Why Do They Say It?* Berkeley: University of California Press, 2000.

Stossel, Thomas P. *Pharmaphobia: How the Conflict of Interest Myth Undermines American Medical Innovation*. Lanham, MD: Rowman & Littlefield, 2015.

INDEX